50 Advances in Biochemical Engineering Biotechnology

Managing Editor: A. Fiechter

Measurement and Control

With contributions by
R. C. Anand, G. H. Dar, G.G. Guilbault,
H. Jung, K. Jung, H.-P. Kleber,
L. C. Lievense, J. H. T. Luong,
A.-L. Nguyen, P. K. Sharma, K. Shimizu,
K. van 't Riet

With 17 Figures and 12 Tables

Springer-Verlag
Berlin Heidelberg NewYork
London Paris Tokyo
Hong Kong Barcelona Budapest

ISBN 978-3-662-14953-9   ISBN 978-3-540-47587-3 (eBook)
DOI 10.1007/978-3-540-47587-3

Library of Congress Catalog Card Number 72-152360

Springer-Verlag Berlin Heidelberg 1993

Originally published by Springer-Verlag Berlin Heidelberg New York in 1993.
Softcover reprint of the hardcover 1st edition 1993
The use of registered names, trademarks, etc. in this publication does not imply, even
in the absence of a specific statement, that such names are exempt from the relevant
protective laws and regulations and therefore free for general use.

Typesetting: Macmillan India Ltd., Bangalore-25

02/3020 - 5 4 3 2 1 0 - Printed on acid-free paper

Attention all "Enzyme Handbook" Users:

A file with the complete volume indexes Vols. 1 through 5 in delimited ASCII format is available for downloading at no charge from the Springer EARN mailbox. Delimited ASCII format can be imported into most databanks.

The file has been compressed using the popular shareware program "PKZIP" (Trademark of PKware INc., PKZIP is available from most BBS and shareware distributors).

This file distributed without any expressed or implied warranty.

To receive this file send an e-mail message to:
SVSERV@DHDSPRI6.BITNET.
The message must be: "GET/ENZHB/ENZ_HB.ZIP".

SPSERV is an automatic data distribution system. It responds to your message. The following commands are available:

HELP	returns a detailed instruction set for the use of SVSERV,
DIR (*name*)	returns a list of files available in the directory "name",
INDEX (*name*)	same as "DIR"
CD <*name*>	changes to directory "name",
SEND <*filename*>	invokes a message with the file "filename"
GET <*filename*>	same as "SEND".

Table of Contents

Genetically Engineered Microorganisms to Rescue Plants from Frost Injury

G. H. Dar[1], R. C. Anand[2] and P. K. Sharma[2]*
[1] Microbiology Section, Division of Plant Pathology,
S.K. University of Agricultural Sciences and Technology,
Shalimar, Srinagar, Kashmir-191 121, India
[2] Department of Microbiology, Haryana Agricultural University,
Hisar-125 004, India

Ice nucleation active bacteria belonging to genera *Pseudomonas*, *Xanthomonas* and *Erwinia* contribute to frost damage to plants by initiating the formation of ice in plants that would otherwise supercool and avoid the damaging ice formation. The biological control of frost injury can be achieved by the application of non-ice nucleation active bacteria to the plant surfaces before they become colonized by Ice⁺ species. *ice* genes have been cloned from *Pseudomonas* and isogenic Ice⁻ derivatives constructed via genetic manipulations. These genetically engineered microorganisms (GEMs) have been released into the environment to control the frost damage. The incidence of frost injury to the plants has, thereby, been reduced by 50–85% during natural frosts. These GEMs do not survive in soil and show no aerial dispersal in the environment.

* To whom correspondence should be addressed

Advances in Biochemical Engineering
Biotechnology, Vol. 50
Managing Editor: A. Fiechter
© Springer-Verlag Berlin Heidelberg 1993

1 Introduction

The use and release of genetically engineered microorganisms (GEMs) into the environment has become a tenable proposition in the last decade. Such organisms include both those produced by recombinant techniques as well as those derived by intensive selection of native species [1]. A variety of uses of GEMs has been identified in crop production and protection, degradation or sequestration of environmental pollutants, extraction of metals from ores, industrial bioconversions and the production of enzymes [2]. Recently, GEMs have been successfully exploited to control frost injury to plants [3]. The ability of some species of Gram negative bacteria and lichens to nucleate the crystallization of water at low temperature demonstrates a unique type of manipulation of the environment by microorganisms. Such microorganisms are major ice nucleating agents found on the leaves and flowers of many plants and initiate much of damage done to crops by frost besides having a role in the rain cycle and in freeze texturing food [4–7]. Much attention has recently been paid to the role of ice nucleating activity in the process of plant damage by *Pseudomonas syringae* since ice formation is a prerequisite for many diseases induced by this bacterium.

The various parameters of Ina (ice nucleation active) bacteria such as their physiology, biochemical and genetical characteristics as well as their ecology have been studied in detail in recent years [8–10]. The molecular basis of ice nucleation has also been worked out. The ice nucleating activity in different microorganisms is conferred by a single structural gene. GEMs have been constructed by manipulating this gene and isogenic mutants (Ice$^-$ or Ina$^-$) are currently being used to control frost injury to plants [11]. In the present article, the bacterial induction of ice nucleation, the extent of frost damage to plants and its control by GEMs under field conditions are discussed.

2 Frost Injury in Plants

Frost damage is one of the major cause of crop loss in temperate and subtropical regions and low temperature has been reported as being the most limiting factor to natural plant distribution [11a]. Frost sensitive plants are distinguished from frost tolerant plants by their relative inability to tolerate ice formation within their tissues. Tolerant plants can do so thereby avoiding the frost damage [12–15] which occurs to sensitive plants between -2 to $-5\,°C$ under natural conditions [16–18]. At these temperatures, ice formed from super-cooled water propagates throughout the tissue system (inter- and intra-cellularly) in such plants and results in frost damage. In the absence of site capable of ice nucleation, the water in the plants can supercool and freezing will not occur until the temperature becomes low enough to allow plant components to catalyze crystallization of supercooled water.

The frost damage is mostly prevalent in temperate regions and to some extent in sub-tropical areas during winter and spring seasons. The severe damage occurs to foliage and flowers in field and horticultural crops. Such damage at the time of budding and flowering in pear, peach, almond and other crops not only reduces yield but also predisposes the development of other diseases such as dieback, canker etc. [8, 9, 19, 20]. Even shoots and stems of trees like pine and eucalyptus and crop plants such as papaya, mango, wheat, potato, tomato etc. become susceptible to such injury. Most of the frost sensitive plants have no mechanism for frost tolerance or for recovery from damage, thus they must be protected from ice formation to avoid frost injury [12, 14, 15].

3 Bacterial Induction of Ice Formation

Although ice melts at 0 °C, water does not necessarily freeze at this temperature. Instead, it can be supercooled to much lower temperatures (as low as -40 °C) before freezing occurs [21, 22]. This is because the initiation of ice crystallization depends on the presence of ice nuclei – the particles of critical size and specific shape that allows formation of an ice lattice around them. In pure water, ice nuclei can be created only by chance orientation of water molecules. However, in the presence of some contaminants in water, freezing occurs at higher temperature. For instance, dust particles can raise the freezing temperature to -10 °C and mineral particles such as silver iodide can cause ice nucleation at -8 °C [23]. However, the members of bacterial genera such as *Pseudomonas, Erwinia, Xanthomonas* and a few lichen of the genera *Rhizoplaca, Xanthoparmelia, Xanthuria* are most active ice nucleating agents [24–29].

Most phyllosphere colonizing organisms show ice nucleation activity at temperatures warmer than -15 °C [3, 30, 31]. However, seven bacterial species namely *Pseudomonas syringae* [24, 32]; *Pseudomonas fluorescens* [3, 26, 38]; *Pseudomonas viridiflava* [3, 38]; *Erwinia herbicola* [33]; *Erwinia ananas* [35, 122]; *Erwinia uredovora* [35] and *Xanthomonas campestris* pv. *translucens* [34] are active in ice nucleation in vitro as well as on the plant surface at temperatures warmer than -1.5 °C or when the temperature falls below zero as may happen in early spring or autumn (Table 1). The plant may supercool but the capacity to supercool is reduced in the presence of Ina bacteria in the phyllosphere which increases the frost injury to the plant. There is direct correlation between frost injury and population of epiphytic Ina bacteria. The Ina bacteria occur in several habitats including water, plant surfaces and in the environment above the plant [24, 36, 37, 84]. Many strains of these colonizers are either epiphytic saprophytes or conditional plant pathogens. The most ubiquitous Ina bacterium *P. syringae*, a pseudomonad pathogenie to liliac, is a common harmless commensal epiphyte for most other plants from widespread geographic locations [6, 24, 38, 39]. Of the 44 pathovars (host range variants) of *P. syringae* known so far, about 50% are active ice nucleators [38–40].

Table 1. List of micro-organisms involved in ice nucleation at temperatures warmer than $-1.5\,°C$ and the corresponding host plants

Sr. No.	Micro-organisms	Host plants	Reference No.
A. Bacteria			
1.	*Pseudomonas syringae*		
	pv. *syringae*	Tomato, beans, corn	38, 51, 112, 117
	pv. *pisi*	Pea	38, 51
	pv. *coronafaciens*	Oats	38
	pv. *lachrymans*	Cucumber	38
	Other epiphytic strains	Cereals, corn, soyabean, beans, potato, tomato, hairy vetch and some other vegetables, clover lilae, strawberries, citrus, coniferous trees, olive, pear, peach, cherry, almond, apricot and other deciduous fruit trees.	5, 18, 19, 32, 33, 86, 110, 111, 112, 118
2.	*P. fluorescens* (epiphytic)	Pine and some forest plants, some cereals, vegetables and fruit crops etc.	3, 38, 61, 76, 116
3.	*P. viridiflava* (epiphytic)	– do –	3, 38, 121, 122
4.	*Erwinia herbicola* (epiphytic)	Corn, cereals, sugarcane, clover, timothy, perennial ryegrass and other grasses, vegetables, fruits and flowers of some deciduous plants etc.	17, 89, 104, 113, 114
5.	*E. ananas* (epiphytic)	Vegetable crops	122
6.	*E. uredovora* (epiphytic)	Strawberry	35
7.	*Xanthomonas campestris* pv. *translucens*	Rice, wheat, corn, oat, various vegetable crops etc.	25, 104
B. Lichens			
1.	*Rhizoplaca*	–	28
2.	*Xanthoparmelia*	–	29
3.	*Xanthuria*	–	29

The ice nucleation frequency (the ratio of the number of ice nuclei to the number of bacterial cells capable of ice nucleation) on plant varies from 10^{-8} to 10^{-1} with an average frequency of 10^{-3} [41–45]. One ice nucleus is sufficient to cause ice formation and thereby frost damage to the entire leaf [43]. Thus, the disease basically abiotic in nature becomes biotic at temperatures warmer than $-1.5\,°C$ when the Ina bacterial population exceeds a threshold number of about $10^3\ g^{-1}$ fresh weight of plant tissue or 10^{12} cfu (colony forming unit) cm^{-2} of leaf surface [44, 46]. Detection of Ina bacteria under laboratory conditions has been reviewed earlier [47]. The Ina bacterial colonies are detected by replica freezing of leaf material on laboratory media at $-5\,°C$, followed by a fine mistspray of ice-nucleation-free water on the surface. The Ina^+ colonies appear as discrete frosty white areas of ice [48–49]. Measurement of the supercooling point of leaves by the tube nucleation assay is also predictive of plant frost sensitivity under field conditions [43, 48].

The rate of nucleation at a particular nucleating site increases drastically over a narrow range of temperatures [21]. It is, therefore, essential to consider a point in this range to know the threshold temperature at and below which the sites become active in nucleation. Each particle that can nucleate ice crystallization has its own characteristic threshold temperature. The ice nucleation sites formed by bacteria vary in their threshold temperature from $- 2.5\,°C$ to $- 14\,°C$. Only an extremely small fraction of cells in the culture show ice nucleation activity at various temperatures above $- 5\,°C$. At temperatures below $- 10\,°C$, essentially all cells exhibit nucleation activity. A plot of the number of cells capable of nucleation versus temperature gives a cumulative nucleus spectrum [21]. On the basis of cumulative nucleus spectra three types of nucleating activity have been differentiated viz. cells active at $- 5\,°C$ or warmer designated as type I; $- 5\,°C$ to $- 8\,°C$ as type II; and $- 8\,°C$ or below as type III activity [33, 50]. Subsequently, this classification was modified on the basis of nucleus spectrum, sensitivity of pH, water miscible solvent and has been designated as class A, B and C type nucleating activity [51]. The class A structure is formed on only a small fraction of cells in the culture, nucleate water at a temperature of $- 4.4\,°C$ and is an effective nucleator of D_2O. A second class of structure called class B is formed on a large proportion of cells, nucleate water between $- 4.8\,°C$ to $- 5.7\,°C$ and is a relatively poor nucleator of supercooled D_2O. The class C structure is formed on almost all the cells and nucleates at $- 7.6\,°C$ or cooler. These three classes are also differentiated by their sensitivity to low concentration of dioxane and dimethyl sulfoxide. The addition of these solvents lowers nucleation activity by 1000-fold or more in class A while in classes B and C, it is lowered by 20–40% and 70–90%, respectively.

4 Genetics of Ice Nucleation

A single gene is responsible for the Ina[+] phenotype in each of the bacterial species. This gene has been designed as *ice* E, *ice* C (*ina* Z) and *ina* W in *E. herbicola*, *P. syringae* and *P. fluorescens*, respectively [52–54]. The Ina proteins from different species show a high degree of homology indicating that these genes have descended from a common ancestral gene.

The involvement of the *ina* gene product in nucleating activity has been shown by deletion, mutagenesis or cloning studies [55–56]. Transfer and expression of this gene into other Gram negative hosts has confirmed that each of the known genes is sufficient for the Ina[+] phenotype. Each *ina* gene contains three orders of internal repetition. The first order repeat is eight codons in length. The two first order repeats make up a second order repeat of 16 codons. A triplet of second order repeat make up the third order repeat of 48 codons [8, 53]. Translation of the consensus of 48 codons repeat gives the following peptide: AGYGST-TAG-SSLIAGYGSTQTAG-S-LT-AGYGSTQTAQ-S-LT.

In most species this repetition is perfect throughout the length of gene while in others, it becomes imperfect. However, deviation from the perfect repetition is similar in each gene suggesting that the pattern of imperfection is functional.

The *ina* genes of different bacterial species encode for a single protein of 150–180 kDa [57]. This protein is found in low quantities, and forms about 0.01% of the total membrane protein [58]. However, this small amount of protein is sufficient to confer Ice$^+$ phenotype against all heterogenic backgrounds tested [53, 55]. The Ina proteins have been identified in Ina$^+$ transformants of *E. coli* using antibodies directed against Ina W protein [59] and a synthetic peptide homologous to part of the predicted amino acids sequence responsible for the nucleation activity [60].

5 Structural Basis for Ice Nucleation in Bacteria

The bacterial ice nucleation phenotype is very sensitive to heat (74 °C), pH, proteases, 2-mercaptoethanol, urea, SDS and heavy metals [26, 61, 62] indicating that a protein is required for ice nucleation activity. Ice nucleation activity of *P. syringae* and *E. herbicola* is also partially sensitive to membrane perturbant compounds such as borate or lectins; and the membrane structure changes associated with lysis by a virulent phage [63]. For a bacterium to nucleate ice formation in external super-cooled water, one would expect that the nucleating structure must be freely exposed and should be exterior to the cell. The bacterial ice nucleating site may be localized in the outer membrane of the source organism [37]. Subsequently, it was found that it may be a membrane-bound protein aggregate [4]. Ice nucleation activity has been enriched by fractionation of bacterial membranes in subcellular particles [4, 37]. Isolated outer membranes of *P. syringae*, *E. herbicola* and *E. coli* HB 101 containing a cloned *ice* gene show ice nucleation activity [55]. Further, membrane vesicles shed off from strains of Ina bacteria also show ice nucleation activity [50]. However, it is not true in *P. syringae* since the product of the *ina* Z gene is localized both in the outer and inner cell membrane.

Bacterial Ina proteins require lipids for their activity as the delipidation of the membrane abolished ice nucleation activity, which was then reconstituted by addition of specific lipids [63, 64]. But, lipopolysaccharides (LPS) that occur exclusively in the outer membrane are not required for ice nucleation activity [58]. The removal of LPS up to 78% did not reduce the ice nucleation activity significantly. The hydrophobic environment supplied by phospholipids greatly favours both the assembly and proper orientation of the membrane protein complex. From the different classes of phospholipids, phosphatidyl inositol (PI) which constitutes 0.1 to 1.0% of total phospholipids in Ice$^+$ bacterial strains is essential for the Ina protein of membranes. The Ice$^-$ *E. coli* strains contained traces of PI that amount to 2–30% of the level found in the Ice$^+$ transformed

E. coli. The relative ice nucleation activity at $-4\,^\circ C$ was proportional to PI content. It was, therefore, concluded from these studies that PI plays an important role in ice nucleation at warm temperatures and is a likely component of ice nucleating sites [65]. However, overproduction of Ina proteins in *E. coli* results in the accumulation of a large number of nucleating sites in the inner membrane. It appears that assembly of nucleating sites requires only common lipids and may take place on the inner membrane prior to transport to the outer membrane. No detectable signal peptide for the translocation of Ina proteins is present. However, the mechanism of translocation of the protein is still not fully understood. It is probable that Ina proteins utilize signal peptides with unusual sequences. The three orders of periodicity in the Ina proteins probably reflect the hierarchy of the three motif of structural repetition. It is possible to construct a model matching the hexagonal symmetry of ice by virtue of the 2×3 organisation of the repeat sequences.

Cell nucleating activity is proportional to the size of the nucleating structure [53, 63]. The target size of the ice nucleation structure is roughly proportional to the ability of cells to nucleate at the warmest temperature [58]. A non-linear but positive relationship between the concentration of the *ina* gene product and the ice nucleation activity of a bacterial culture has been reported [66]. Larger aggregates of interacting ice nucleation proteins are required at warmer temperatures. However, there is no evidence of chemical heterogeneity between structures active at different temperatures. The cross-sectional area of a nucleation site (i.e. the size of target it represents) can be inferred from the dosage of γ-radiation required to inactivate it. Assuming that a single hit is sufficient for inactivation of an Ina site, it has been concluded that nucleation sites, with a $-12\,^\circ C$ threshold possess an average molecular size of approximately 150 kDa and a nucleation at $-3\,^\circ C$ required a structure with an apparent size of 700 kDa [64]. Nucleating sites with a higher threshold required smaller radiation doses for inactivation which results in the estimation of a much larger molecular size (19 000 kDa) for nucleation sites with a threshold above $-3\,^\circ C$. However, the large size estimated for nucleation sites with a $-3\,^\circ C$ threshold from theoretical predictions demonstrate that some bacterial ice nucleating sites are considerably larger than a single molecule of Ina protein. A plausible model is that of co-operative homo-aggregates of Ina proteins, embedded in and stabilized by a lipid membrane [8]. This model is supported by the fact that nuclei which are active at warmer temperatures are assembled more slowly [67].

6 Construction of Ice⁻ GEMs

Genomic libraries of *P. syringae* and *E. herbicola* have been constructed. This greatly facilitated the manipulation of genes responsible for the Ice⁺ phenotype. Expression of cloned *ice* genes have been observed in the heterogenous Gram negative bacterium *E. coli*. Recombinants of *E. coli* strains exhibiting the Ina$^+$ phenotype are being used in studies for understanding various parameters such as physiological, biochemical characteristics as well as its genetics [51, 66, 68–70]. The manipulation of a single gene responsible for Ina$^+$ phenotype by deletion, mutation or replacement has facilitated the construction of Ina$^-$ or Ice$^-$ GEMs which are defective in ice nucleation. Various methods have been

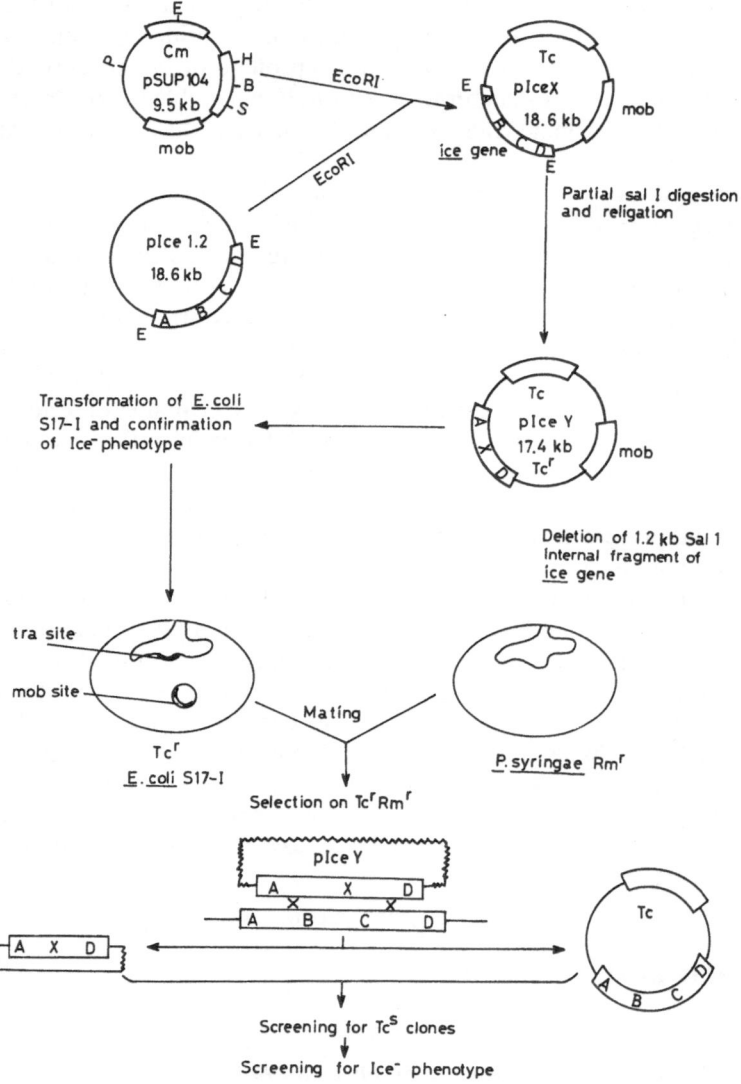

Fig. 1. Construction of Ice$^-$ GEMs by deletion of an internal fragment of the *ice* gene

employed for the construction of Ice$^-$ GEMs e.g. deletion of the *ice* gene; random mutagenesis; and site-directed mutagenesis [68, 71–74].

6.1 Deletion of Ice Gene

Deletion of the *ina* Z gene of *P. syringae* strain 5203 and the *ina* W gene of *P. fluorescens* strain MS 165 has been generated in vitro by using site specific restriction enzyme such as *Sal* I. DNA segments containing an internal deletion of about 1.2 kb (kilobases) in the *ina* Z gene of *P. syringae* were ligated and *E. coli* cells were transformed with this plasmid. Transformed *E. coli* cells showed Ice$^-$ phenotype. This plasmid was transferred from *E. coli* to *Pseudomonas* and the clones showing a double crossover in which the active *ina* Z gene is replaced by the *ina* Z gene with deletion were isolated. All the deletion mutants were of the Ice$^-$ phenotype. The double crossover events occur with low frequency. The frequency of double crossovers could be increased using the sucrose sensitivity of the *sac*BR cassette of *Bacillus subtilis*. The use of the integrative vector facilitates the double crossover. Use of the *sac* BR (sucrose sensitive gene) that increases the frequency of double crossovers provides single step methods for screening of Ice$^-$ mutants [75]. Some of the Ice$^-$ mutants constructed are RGP 36, Cit 7 del 16, TLP 2 del 1 of *P. syringae* and GJP 13b of *P. fluorescens* [76–78] (Fig. 1).

Deletion of part or all of the coding sequence of 48 codons, upstream of the repetitive segment, causes complete loss of nucleating sites with a warmer threshold but only partial loss of those with a colder threshold. In contrast, deletions in the coding sequence downstream from the repeat are shown to be essential for all types of nucleating sites [79]. It is the deletion of the correct fragment which gives Ice$^-$ GEMs.

Another way is to disrupt an active *ice* C gene with an inactive *ice* C gene which lacks both the termini. An *ice* C gene lacking both the termini is cloned into an integrative vector which is unable to replicate in *P. syringae*. Transfer of this plasmid from *E. coli* to *P. syringae* in a single crossover between homologous *ice* C genes on a plasmid as well as on the chromosome disrupt both *ice* C genes; one lacking the beginning sequence while the other lacks its ending sequence [32, 64, 65].

6.2 Random Mutagenesis

Ice$^-$ GEMs from *P. syringae* have been isolated by random *Tn*-5 or chemical mutagenesis. Suicidal plasmids or phages carrying *Tn*-5 can be used to isolate Ice$^-$ mutants. Some of the suicidal plasmids pSUP 2021, pGS 9 are widely used for these types of experiments (Fig. 2) [80, 81]. Ice$^-$ GEMs have been isolated earlier by *Tn*-5 mutagenesis [68]. Chemical mutagens such as EMS (ethyl methyl sulfonate) have been successfully used to construct Ice$^-$ mutants [52].

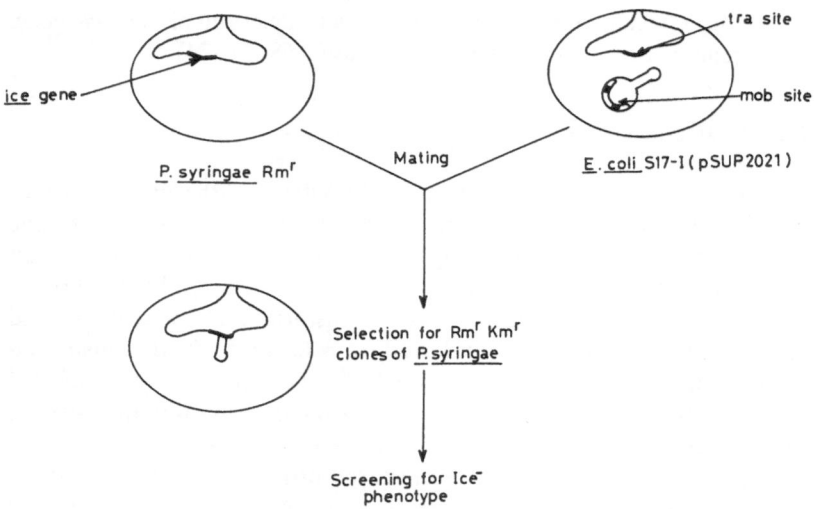

Fig. 2. Construction of Ice⁻ GEMs by random *Tn-5* mutagenesis

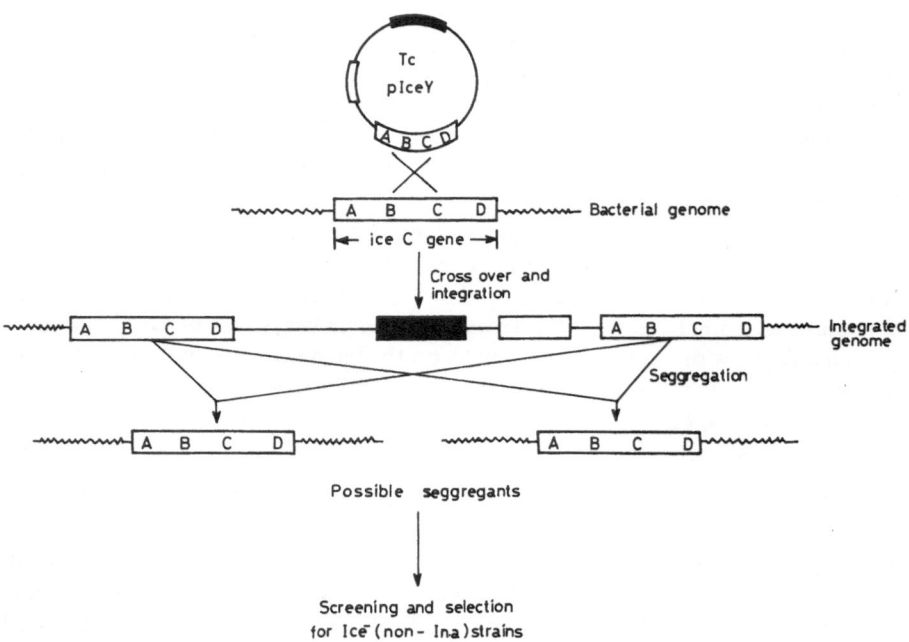

Fig. 3. Gene replacement involving integration of an inactive modified gene (ABCD) carried by a plasmid, into the genome of *Pseudomonas* having an active *ice* C gene (ABCD) by single reciprocal recombination. The segregant *Pseudomonas* having inactive modified genes (ABCD) are then screened

6.3 Site Directed Mutagenesis

A useful technique has been described for creating transposon insertion mutations in a selected region of the genomic DNA of a Gram negative bacteria [74]. In this procedure, termed as marker exchange, transposon insertions are

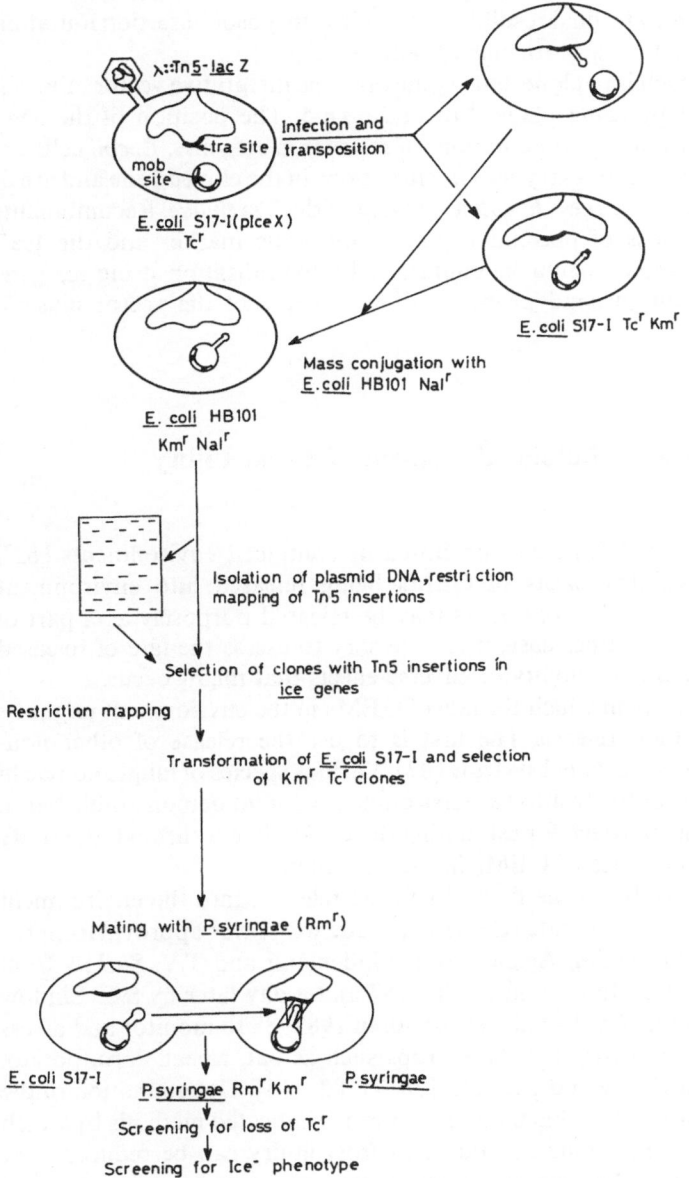

Fig. 4. Construction of Ice⁻ GEMs by site directed mutagenesis using λ :: *Tn-5 Lac* Z phage [119]

obtained within a selected cloned DNA fragment, located on a broad host range plasmid vector in *E. coli*. The position of the insertions can be determined by restriction enzyme analysis. The cloned DNA is introduced into the parental organism and selection is made for recombinants which replace the wild genomic allele with the plasmid borne transposon inserted allele. The latter step involves introducing a second incompatible plasmid to affect the elimination of the original plasmid. The final result is a genomic transposon insertion mutation which can be scored for a particular phenotype (Fig. 3).

Another approach is to clone an *ice* gene on some integrative vector in *E. coli* and mutagenise it by using phage Mu or λ :: *Tn*-5. The position of the *Tn*-5 insertions is mapped using combinations of restriction enzymes. *E. coli* cells are transformed with plasmids carrying *Tn*-5 insertions in the cloned gene and are of the *Ice*⁻ phenotype. The Ice⁻ *E. coli* are mated with *P. syringae*. Recombinants are tested for the loss of plasmid using an antibiotic marker and the Ice⁻ phenotype. The mutants could be confirmed by hybridisation using *ice* gene probes and a vector plasmid probe to show the loss of the vector plasmid (Fig. 4).

7 Role of GEMs in Biological Control of Frost Injury

The proposed uses of GEMs are for closed or contained environments [82]. However, the possibility exists of GEMs being released into environment accidently and several types of GEMs may be released purposely as a part of management plans. In either case, it is necessary to assess the fate of released GEMs and reduce the possibility of adverse effects that might occur.

There are two ways in which the fate of GEMs in the environment might be assessed prior to their release. The first is to use the release of other non-indigenous organisms as model systems [82]. Various species of fungi and bacilli have been released routinely into the environment without demonstrable harm. A more compelling approach for estimating the GEMs fate is through the study of survival and persistence of GEMs in the environment.

The distinction of being the first ever GEM released into the environment goes to *P. syringae*. Its deliberate release was made with prior approval from the US Environment Protection Agency by J. Lindemann and T.V. Suslow from Wisconsin University, Madison in April, 1987 and a day later by S.E. Lindow from California University. Initial tests began in 1987–88 to monitor and assess their potential as biocontrol agents in crops such as oat, wheat, corn, potato, tomato, beans, strawberry and pear etc. [50, 77, 82, 115]. Determination of the population dynamics of Ice⁺ bacteria on plants has indicated methods by which these species can be manipulated and plant frost injury can be reduced. The population size of Ice⁺ bacteria on many plants undergoes seasonal variation [18, 45, 85]. Only 0.01–40.0% of the total bacteria on the plant surface are Ice⁺

strains that are involved directly in plant freeze injury [3]. Thus Ice$^-$ and Ice$^+$ bacteria co-exist on many plants in natural habitats and competition between such bacterial species appears likely. However, the degree of natural competition is generally inadequate to prevent the development of a large population of Ice$^+$ bacteria on many plants. The maximum epiphytic bacterial population size on plants is between 10^7 and 10^8 cells per g fresh weight. While the total population size of *P. syringae* on plants treated with Ice$^-$ mutants of this species approached this number, the population of Ice$^+$ strains among population decreased with increasing population size of Ice$^-$ *P. syringae* [46]. Thus a limited number of habitat sites on leaves or other resources may limit the population size of Ice$^+$ bacteria.

It is essential to spray a large volume of Ice$^-$ GEMs such as *P. syringae* (Cit 7 del 7b or TLP2 del 1) or *P. fluorescens* in field crops [83]. Two sprays with a 15–20 day interval have been found effective to achieve significant protection against frost injury. The reduction in frost damage by such sprays has been 30–90% [86, 87]. Ice$^-$ GEMs have shown good colonizing and survival ability on the plant surface and these compete well with native Ice$^+$ strains [43, 68, 88]. This competition appears sufficient to account for the exclusion of Ice$^+$ bacteria by Ice$^-$ mutants of *P. syringae*. The spray applications have only been successful prior to the colonization of emerging leaves by Ice$^+$ strains since Ice$^-$ mutants are unable to displace the pre-existing population [25, 89, 90]. Thus the competition on the plant surface is principally dependent on the relative population size not on the genotype of the bacterial strain applied [34, 86, 91].

The antagonistic interaction between Ina$^+$ and non-Ina bacteria on the phyllosphere, which is due to competition for limiting environmental resources like nutrients or specific habitable sites, may restrict the colonization of leaves by Ina$^+$ bacteria [26, 76, 86, 87, 92–101]. Antagonistic Ice$^-$ bacteria have been established on plants by application to leaves or to seeds of plants before they are colonized by the ice$^+$ bacteria [24, 37, 44, 46, 86]. The population size of ice$^+$ bacteria have been reduced from 10 to 500-fold on plants treated with antagonistic bacteria when compared to untreated plants under field conditions [4, 37, 44]. Many naturally occurring Ice$^-$ bacteria that are antagonistic to Ice$^+$ bacteria on plants produce antibiotic-like compounds in culture [102, 103]. But, antagonistic mutants of parental strains which were inhibitory to *P. syringae* in culture did not differ from their respective parental strains in controlling *P. syringae* population on the leaf surface [102]. Ice$^-$ mutants of *P. syringae* induced by chemical mutagenesis prevented the growth of isogenic Ice$^-$ strains or other Ice$^+$ *P. syringae* on the plant surface [46, 102]. However, the mechanism of competition between Ice$^+$ and Ice$^-$ strains e.g. major nutrients and physical sites for which competition exists, are still unclear [74–79, 86, 92–95].

A tricky and novel approach to achieve successful prevention of frost injury in crops with already well established Ina$^+$ bacteria, is represented by the use of some specific bacteriophages along with Ina bacterial spray application. The Ina protein is the specific receptor site for these bacteriophages and thereby, they destroy the Ice$^+$ strains of bacteria. Thus the combined spray of Ice$^-$ bacterial

strain and bacteriophages creates a favourable environment for the estab-
lishment of the Ice⁻ strains [61].

The application of a larger bacterial inoculum on the plant surface generally
does not exceed an epiphytic population size of 10^7 cells per g [37, 44, 73]. The
non-Ina (Ice⁻) bacterial species applied were recovered from the plant surface
over a period of one day to more than two months. The aerosol studies have
revealed that 99.99% of the sprayed bacteria settle directly on plant and soil
[104, 105]. Furthermore, the plant material is essential for their survival [44].
The knowledge of a potential competitive advantage of one strain over other
would be useful in predicting the outcome of large scale releases of non-Ina
bacterial strains for frost control in crop and horticultural plants.

8 Risk Assessments

The major concern about the introduction of GEMs into the soil and other
natural environments is their potential ecological impact especially on homeo-
stasis of these environments. Thus development of sensitive techniques to
monitor growth, survival and gene transfer by GEMs in these environments is
necessary before meaningful ecological data can be obtained. The probable
behaviour of GEMs prior to release are assessed to microcosm studies to allay
legitimate public concerns. The increased interest in the use of GEMs has
emphasised the need for an accurate, sensitive, selective and objective methodology for evaluating the risk involved in their release into the environment [46,
106]. Risk assessment of GEMs sprayed on crops include the determination of
off-site disposal and deposition, horizontal and vertical gene transfer, interaction gene transfers, interaction with beneficial residents in soil, impact on
plant and human health and ecological disturbances [103, 105, 107, 108].
Microcomputer stimulation has been used in disposal and deposition studies to
incorporate particle size distribution, wind speed direction of turbulence, temperature, sedimentation rate and mortality [109].

In fields, the release, detection and monitoring of GEMs, is of foremost
importance. Conventional cultural methods have failed to ascertain the fate of
the released organisms since they provide information of only 10% of the
bacteria under study in the soil habitat. A large number of dormant and non-
culturable cells are not detected or monitored by this method. Moreover,
microbial activity, potential for survival and genetic transfers cannot be ana-
lysed by such methods. The population of Ice⁻ mutants of *P. syringae* are
determined by growing them on a medium specific for the Ice⁻ construct (SSM
medium) [68]. The identical colonies resembling Ice⁻ mutant strains on selective media are verified by dot blots or Southern hybridization with P^{32}-labeled
oligonucleotide probes (20–25 bp) specific to these strains [9, 85].

However, recombination may occur in the soil or may be transferred to other micro-organisms by transformation. The polymerase chain reaction (PCR) method has allowed the amplification of specific DNA sequences used as probes. A cyclic series of primer-directed enzymatic polymerization steps alternated with denaturation of the complementary strand permits the amplification of the desired targeted nucleic acid to a measurable amount within a few hours. The PCR method when coupled with direct extraction of DNA from environmental samples has increased the sensitivity of detection to as few as one cell per g of soil [3, 18, 45, 46].

Species of monoclonal antibodies that are specific to host or recombinant plasmids are also used for this purpose. Antibodies against a region of flagellar protein of *P. syringae* recognise the flagellar membrane of released bacteria. Immunofluorescence techniques, nucleic acid sequence analysis, plasmid and protein analysis or antibiotic resistant markers have also been employed for detection of GEMs [18, 45, 85].

The dispersal of Ice⁻ GEMs studied by different methods have revealed that there was no aerial dispersal of sprayed GEMs. The Ice⁻ GEMs were not detected more than 10–20 m outside the sprayed field. They were not recovered from source vegetation or surrounding water. The inoculated strain survival was very poor in soil. No GEM was detected after one week of spray. But, a large population of sprayed GEMs was present on target plants. These observations indicate the suitability of Ice⁻ mutants for the control of frost injury.

9 Conclusions and Future Prospects

Seven bacterial species have been identified as most active ice nucleation bacteria on the surface of healthy plants. These species can contribute to frost damage to plants by initiating ice formation in plants. The biological control of frost injury has been achieved to the extent of 50 to 85% by spraying large volumes of Ice⁻ genetically constructed micro-organisms applied to a potato field. Application of Ice⁻ GEMs reduces the population size of Ice⁺ bacteria by 30 to 180-fold on inoculated plants. However, there is 1000-fold decrease in the Ice⁺ population of *P. syringae* when isogenic Ice⁻ GEMs are used. The larger reduction in the population of isogenic strains suggest that some specificity is involved in competition among the bacteria on the plant surfaces. Therefore, it is essential to construct more Ice⁻ isogenic GEMs to control specific types of Ice⁺ bacteria as naturally occurring bacteria may not be able to compete effectively with all other Ice⁺ bacteria.

Information is being generated on the frequency of gene exchange among the bacteria in natural and managed environments. Gene rearrangements, deletions and transfers are the common phenomena which would affect the fate of engineered DNA sequences. Although no horizontal and vertical transfer of the Ice⁻ strain genome to other bacterial species has been reported yet, such a possibility cannot be ruled out in a natural system.

For commercialisation of Ice⁻ GEMs, more information is needed to understand the following points:

– the role of normal Ina-bacteria;
– the specificity of interaction of these bacteria on phyllosphere and the possible environmental consequences of modifying the normal ecological relationship between Ina-bacteria and the plant upon which they reside;
– temporal response of GEMs to environmental fluctuations;
– the possible influence of these Ina bacteria on other unknown organisms;
– environmental resource for bacterial competition on plant surfaces;
– interaction of host plant species and varietal response;
– whether Ice⁺ and Ice⁻ bacterial species co-exist on the same plant or not (because effective control can only be achieved if co-existence of such dissimilar strains occurs);
– horizontal and vertical gene transfer;
– complete genome and gene product analysis and characterization (of the organisms selected);
– development of appropriate screening and monitoring techniques specific to the strain used.

The improvement in the reliability of these GEMs as inoculants will depend upon the greater understanding of these aspects.

10 References

1. Cairns J, Pratt JR (1986) Aquat Toxicol Environ Fate 9: 207
2. Lindow SE, Panopoulos NJ, McFarland BL (1989) Science 244: 1300
3. Lindow SE, Panopoulos NJ (1988) In: Sussman M, Collins CH, Skinner FA, Stewart-Tull DE (eds) The release of genetically engineered micro-organisms, Academic, London, p 121
4. Lindow SE (1983) Plant Disease 67: 327
5. Gross DC, Cody YS, Proebsting EL, Radamaker GK, Spotts RA (1984) Phytopathology 74: 241
6. Vigouroux A (1989) Plant Disease 73: 854
7. Arai S, Watanabe M (1986) Agric Biol Chem 50: 169
8. Warren G, Wolber P (1991) Mol Microbiol 5: 239
9. Zhao J, Orser CS (1990) Mol Gen Genet 233: 163
10. Watanabe M, Emori Y, Arai S, Watabe S (1990) Mol Microbiol 4: 1871
11. Lindow SE, Panopoulos NJ, Pierce C, Andersen G, Lim G (1988) Phytopathology 78: 1552
11a. Parker J (1963) Bot Rev 29: 124
12. Burke JJ, Gusta LA, Quamme HA, Weiser CJ, Li PH (1976) Ann Rev Plant Physiol 27: 507
13. Chandler WH (1958) Proc Ann Soc Hortic Sci 64: 552
14. Levitt J (1972) In: Responses of plant to environmental stress. Academic, New York, p 306
15. Mazur P (1969) Ann Rev Plant Physiol 20: 419
16. Schnell RC, Vali G (1972) Nature 236: 163

17. Lindow SE, Arny DC, Upper CD (1978) Appl Environ Microbiol 36: 831
18. Gross DC, Cody YS, Proebsting EL, Radamaker GK, Spotts RA (1983) Appl Environ Microbiol 46: 1370
19. Gross DC, Proebsting EL, Jr., Andrews GK (1984b) J Am Soc Hortic Sci 109: 375
20. Montesinos E. Vilardell (1991) Phytopathology 81: 113
21. Vali G, Stansbury EJ (1966) Can J Physiol 44: 477
22. Vali G (1971) J Atoms Sci 28: 402
23. Zettelmeyer AC, Teheurekdjian N, Chessick JJ (1961) Nature 192: 653
24. Maki LR, Galyan EL, Chang-Chien M, Caldwell DR (1974) Appl Microbiol 28: 456
25. Lindow SE (1978) In: Li PH, Sakai A (eds) Plant cold hardiness and freezing stress mechanism and crop improvement. Academic, New York, p 395
26. Maki LR, Willoughby KJ (1978) J Appl Meteorol 17: 1049
27. Anderson SA, Ashworth EN (1986) Plant Physiol 80: 956
28. Kieft TL (1988) Appl Environ Microbiol 54: 1678
29. Kieft TL, Ahmadjian V (1989) Lichenologist 21: 355
30. Anderson JA, Buchanan DW, Stall RE, Hall CB (1982) J Am Soc Hortic Sci 107: 123
31. Anderson JA, Buchanan DW, Stall RE (1984) J Am Soc Hortic Sci 110: 401
32. Arny DC, Lindow SE, Upper CD (1976) Nature 262: 282
33. Yankofsky SA, Levin Z, Moshe A (1981) Curr Microbiol 5: 213
34. Lim HK, Orser C, Lindow SE, Sands DC (1987) Plant Disease 71: 994
35. Obata H, Takinamek, Tanishita J, Hasegawa Y, Kawate S, Tokuyama T, Ueno T (1990) Agric Biol Chem 54: 1990
36. Crosse JE (1971) In: Preece TF, Dickison CH (eds) Ecology of leaf surface micro-organisms, Academic, London, p 282
37. Lindow SE (1983) Ann Rev Phytopath 21: 363
38. Paulin JP, Luisetti J (eds) (1978) Proceedings of the 4th International conference on plant pathogenic bacteria, vol 2, Angers, France, Inst Nat Rech Agron, Beaucouze, 1978, p 717
39. Dye DW, Bradbury JF, Goto M, Hayward AC, Lelliott RA, Schroth MN (1980) Rev Plant Pathol 59: 153
40. Newton D, Hayward AC (1986) Plant Pathol 15: 71
41. Lindow SE (1986) In: Day PR (ed) Biotechnology and crop improvement and protection, BCPC Monograph No. 34, British Crop Protection Council, Cambridge, U.K., p 185
42. Hirano SS, Maher EA, Kelman A, Upper CD (eds) (1978) Proceedings of the 4th International conference on plant pathogenic bacteria, vol 2, Angers, France, Inst Nat Rech Agron, Beaucouze, 1978, p 717
43. Lindow SE, Arny DC, Barchet WR, Upper CD (1978) In: Li PH, Sakai A (eds) Plant cold hardiness and freezing stress mechanism and crop improvement, Academic, New York, p 249
44. Lindow SE, Hirano SS, Arny DC, Upper CD (1982) Plant Physiol 70: 1090
45. Lindow SE, Connell JH (1984) J Am Soc Hortic Sci, 109: 48
46. Lindow SE (1987) Appl Environ Microbiol 53: 2520
47. Hirano SS, Upper CD (1986) Methods Enzymol 127: 730
48. Hirano SS, Baker LS, Upper CD (1985) Plant Physiol 77: 259
49. Lindow SE (1982) In: Mount MS, Lacy GH (eds) Phytopathogenic prokaryotes, vol 1. Academic, New York, p 335
50. Phelps P, Gidding TH, Prochoda M, Fall R (1986) J Bacteriol 1967: 496
51. Turner MA, Arellano F, Kozloff LM (1990) J Bacteriol 172: 2521
52. Orser C, Staskawicz BJ, Panopoulos NJ, Dahlbeck D, Lindow SE (1985) J Bacteriol 164: 359
53. Green RL, Warren GJ (1985) Nature 317: 645
54. Corroto L, Wolber PK, Warren GJ (1986) EMBO J 5: 231
54a. Warren G, Corotto L (1989) Gene 85: 239
55. Buttner MP, Amy PS (1989) Appl Environ Microbiol 55: 1690
56. Warren GJ, Lindemann J, Suslow TV, Green RL (1987) Ice deficient bacteria as frost protection agents. In: LeBaron HM, Mumma RO, Honeycutt RC, Duesing JH (eds). Biotechnology in agricultural chemistry. American Chemical Society, Washington DC, p 215
57. Wolber PK, Deininger CA, Southworth MW, Vandekerckhove J, Van Montagu M, Warren GJ (1986) Proc Natl Acad Sci (USA) 83: 1194
58. Govindarajan AG, Lindow SE (1988) Proc Natl Acad Sci (USA) 85: 771
59. Deininger CA, Mueller GM, Wolber PK (1988) J Bacteriol 170: 669
60. Kaku S (1975) Cryobiology 12: 154

61. Kozloff LM, Schofield MA, Lute M (1983) J Bacteriol 153: 222
62. Sprang ML, Lindow SE (1981) Phytopathology 71: 256
63. Kozloff LM, Lute M, Westaway D (1984) Science 226: 845
64. Govindarajan AG, Lindow SE (1988) J Biol Chem 263: 9333
65. Kozloff LM, Turner MA, Arellano F, Lute M (1991) J Bacteriol 173: 2053
66. Southworth MW, Wolber PK, Warren GJ (1988) J Biol Chem 263: 15211
67. Jasson JT, Holben WE, Tiedje JM (1989) App Environ Microbiol 55: 3032
68. Lindow SE (1988a) In: Megusar F (ed) Microbiol ecology. Am Soc Microbiol Washington, p 187
69. Lindow SE (1986) Phytopathology 76: 1194
70. Ashworth EN (1990) Plant Physiol 92: 718
71. Pooley L, Brown TA (1991) FEMS Microbiol Lett 77: 229
72. Orser CS, Staskawicz BJ, Loper J, Panopoulos NJ, Dahlbeck D, Lindow SE, Schroth MM (1983) In: Pühler A (ed) Molecular genetics of the bacteria-plant interaction. Springer Verlag, Berlin Heidelberg New York, p 353
73. Orser CS, Lotstein R, Lathue E, Wills DK, Panopoulos NJ, Lindow SE (1984) Phytopathology 74: 798
74. Ruvkun GB, Ausubel FM (1981) Nature (London) 289: 858
75. Ried J, Collmer A (1987) Gene 57: 239
76. Lindow SE (1985a) In: Windel CE, Lindow SE (eds) Biological control on the phylloplane. Am Phytopath Soc St Pauli, p 83
77. Lindow SE (1988c) Megusar F, Gantor M (eds) In: Microbial ecology, Slovene Society for Microbiology, Ljuvljana, p 509
78. Warren GJ, Lindemann J, Suslow TV, Green RL (1987) In: LeBaron HM, Mumma RO, Honeycutt RC, Duesing JH (eds) ACS symposium series No. 334, Biotechnology in agricultural chemistry. American Chemical Society, Washington DC, p 215
79. Holben WE, Jasson JK, Chelm BK, Tiedje JM (1988) Appl Environ Microbiol 54: 703
80. Simon R, Priefer U, Pühler A (1983) Biotechnology 1: 784
81. Selvaraj G, Iyer VN (1985) Plasmid 13: 70
82. Brill WJ (1985) Science 227: 381
83. Supkoff DM, Bezark LG, Opgenorth D (1989) In: Monitoring of winter 1987 first release of genetically engineered bacteria in Contra Costa Country. Report BC-88-1, Dept of Food and Agric Sacramento, California
84. Lindow SE, Arny DC, Upper CD (1978) Appl Environ Microbiol 36: 831
85. Hirano SS, Upper CD (1983) Ann Rev Phytopath 21: 243
86. Lindow SE (1985) In: Halvorson OH, Pramer D, Rogul M (eds) Engineered organisms in the environment. Am Soc Microbiol, Washington DC, p 23
87. Lindow SE (1988) Phytopathology 76: 1331
88. Lindemann J, Suslow TV (1987) Phytopathology 77: 882
89. Lindow SE, Arny DC, Upper CD (1983) Phytopathology 73: 1097
90. Lindow SE, Arny DC, Upper CD (1983) Phytopathology 73: 1102
91. Lim HK, Orser C, Lindow SE, Sands DC (1987) Plant Disease 71: 994
92. Abe K, Watabe S, Emori Y, Watanabe M, Arai S (1989) FEBS Lett 258: 297
93. Warren G, Corotto L, Wolber PK (1986) Nucleic Acids Res 14: 8047
94. Morel JL, Bitton G, Chaudhary G, Awong J (1989) Curr Microbiol 18: 355
95. Lindow SE, Wills DK, Panopoulos NJ (1987) Phytopathology 77: 1768
96. Laben C (1965) Ann Rev Phytopathol 3: 209
97. Laben C, Daft GC (1965) Phytopathology 55: 760
98. Laben C, Schroth MN, Hildebrand DC (1970) Phytopathology 60: 677
99. Ki-Chung Kim, Young- Cheolkim, Baik- HoCho (1989) Phytopathology 79: 275
100. Beer SV, Norelli JL, Rundle JR, Hodges SS, Palmer JR, Stein JI, Aldwinckle HS (1980) Phytopathology 70: 459
101. Chakravarti BP, Laben C, Daft GC (1972) Can J Microbiol 18: 696
102. Lindow SE (1985) In: Fokkema N (ed) Microbiology of the phylloplane, Cambridge University Press, London, p 293
103. Lindow SE, Knudsen GR, Seidler RJ, Watter MV, Lambou VM, Arny PS, Schmedding D, Prince V, Herns S (1988) Appl Environ Microbiol 54: 1557
104. Lindow SE, Arny DC, Upper CD (1978) Phytopathology 68: 523
105. Lindow SE, Knudsen GR, Seidler RJ, Walter MV, Lambou VM, Arny PS, Schmedding D, Prince V, Herns S (1988) Appl Environ Microbiol 54: 2281

106. Jones DA, Rijder MH, Clare BG, Ferrand SK, Kerr A (1988) Mol Gen Genet 212: 207
107. Suslow TV (1989) Phytopathology 79: 1151
108. Choudhary G, Awong J (1989) Curr Microbiol 18: 355
109. Knudsen GR (1989) Appl Environ Microbiol 55: 2641
110. Klement Z, Rozsnyay DS, Balo E, Panczel M, Prileszky G (1984) Physiol Plant Pathol 24: 237
111. Ashworth EN, Anderson JA, Davis GA (1985) J Am Soc Hortic Sci 110: 287
112. Lindemann J, Joe L, Moayeri A (1985) Phytopathology 75: 1361
113. Lindow SE, Staskawicz BJ (1981) Phytopathology 71: 237
114. Lindow SE, Arny DC, Upper CD (1982) Plant Physiol 70: 1084
115. Lindow SE (1985) In: Fokkema N, Van Den Heuvel J (eds) Microbiology of phyllosphere, Cambridge University Press, London, p 293
116. Hirano SS, Upper CD (1986) In: Fokkema NJ, Van Den Heuvel J (eds) Microbiology of phyllosphere, Cambridge University Press, London, p 235
117. Hirano SS, Nordheim EV, Arny DC, Upper CD (1982) Appl Environ Microbiol 44: 695
118. Makino T (1982) Ann Phytopathol Soc Japan 48: 452
119. Keller M, Müller P, Simon R, Pühler A (1988) Mol Plant Microbe Interact 1: 267
120. Obata H, Saeki Y, Tanishita J, Tokuyama T, Higashi Y (1987) Agric Biol Chem 51: 1761
121. Obata H, Nakai T, Tanishita J, Tokuyama T (1989) J Ferment Bioeng 67: 143
122. Goto M, Haung BL, Makino T, Goto T, Inaba T (1988) Ann Phytopathol Soc Jpn 54: 189

Synthesis of L-Carnitine by Microorganisms and Isolated Enzymes

H. Jung, K. Jung and H.-P. Kleber
Biochemie, Fachbereich Biowissenschaften, Universität Leipzig,
Talstraße 33, O-7010 Leipzig, FRG

L-Carnitine, a quaternary ammonium compound, plays an important role in β-oxidation of fatty acids in mammals. The increasing demand for this compound in medicine has led to the development of numerous procedures for L-carnitine production. This review discusses the possibilities of microbial and enzymatical synthesis of L-carnitine and gives an overview on the pathways of L-carnitine metabolism and related enzymes in microorganisms.

Advances in Biochemical Engineering
Biotechnology, Vol. 50
Managing Editor: A. Fiechter
© Springer-Verlag Berlin Heidelberg 1993

1 Introduction

L-Carnitine ((R)-3-hydroxy-4-N-trimethylaminobutyrate) is ubiquitous in cells of higher animals. Furthermore, it has been found in higher plants and some microorganisms [1]. L-Carnitine was first isolated from meat extract in 1905 [2, 3] and its structure was established by chemical synthesis in 1927 [4]. Twenty five years later L-carnitine was shown to be an essential growth factor for the meal worm, *Tenebrio molitor*, and hence it was called vitamin B_T [5]. Further investigations by Fritz [6] and Bremer [1] demonstrated a function for L-carnitine in the β-oxidation of fatty acids. It was shown that this quaternary ammonium compound is essential for the transport of long-chain fatty acids through the inner mitochondrial membrane of mammals. Furthermore, L-carnitine may play a role in shuttling short-chain and medium-chain acyl residues out of peroxisomes in tissues that chain-shorten long-chain fatty acids via peroxisomal β-oxidation [1]. L-Carnitine facilitates the oxidation of branched chain α-keto acids and may modulate the acylCoA/CoASH ratio [7].

One source of L-carnitine of the healthy mammalian organism is the uptake from food [8, 9]. In addition, L-carnitine may be synthesized from the two essential amino acids lysine and methionine [10, 11]. In *Neurospora crassa* free lysine is methylated with S-adenosylmethionine as methyl donor [12, 13]. In animals, however, 6-N-trimethyllysine is formed by methylation of lysine which is bound in proteins such as histones, myosin and actin [14]. Further intermediates in L-carnitine biosynthesis are 3-hydroxy-6-N-trimethyllysine [15, 16], 4-N-trimethylaminobutyraldehyde [15, 16], 4-N-trimethylaminobutyrate (γ-butyrobetaine) [17, 18]. In humans, the last step, the hydroxylation of γ-butyrobetaine, occurs in liver, kidney and brain but not in cardiac or skeletal muscle [16, 19].

In 1973, Engel and Angelini [20] reported for the first time patients with carnitine deficiency as a progressive skeletal myopathy with extensive lipid vacuoles evident on muscle biopsy. Among the three clinical types of carnitine deficiency – myopathic, systemic, and mixed forms – the two latter may be treated by application of L-carnitine with some success [21, 22]. Furthermore, L-carnitine is used in replacement therapy of haemodialysis patients as well as in prophylaxis and therapy of various heart diseases [23]. Recently there has also been increasing interest in the protective effect of L-propionylcarnitine on the ischemic heart [24].

Cheap, racemic D,L-carnitine cannot longer be applied in clinics because the D-enantiomer is not as harmless as was assumed in the past [25]. Thus, D-carnitine is bound and transported by the active L-carnitine transport system of cell membranes thereby diminishing L-carnitine within the cells and inhibiting L-carnitine specific reactions.

Beside its application in medicine, L-carnitine may be used for stimulation of the growth of yeasts [26, 27], obligatory and facultative methylotrophic bacteria

[28] as well as bacteria which are grown on *n*-alkanes (*Acinetobacter* and *Pseudomonas* species) [29]. Furthermore, L-carnitine is a suitable supplement for the serum-free cultivation of murine hybridoma cells. It stimulates the production of antibodies by these cells [30].

The growing demand for L-carnitine, particularly in medicine, has caused a world-wide search for ways of synthesizing this betaine in an optically pure form. Some chemical procedures are based on the resolution of racemic carnitine or its precursors via their diastereomers by means of optical active acids [31–35]. Today, most L-carnitine (approximately 200 t per year) is produced by application of these methods. Another chemical possibility of obtaining L-carnitine involves chemical multi-step synthesis starting from chiral precursors like, e.g. L-ascorbic acid [36], D-mannitol [37], (*R*)-4-chloro-3-hydroxybutyrate [38, 39], 4-hydroxy-L-proline [40] or malic acid [41].

This review discusses methods of L-carnitine production by microorganisms and enzymes. The knowledge of carnitine metabolizing enzymes related to these methods is summarized.

2 Resolution of Racemic Carnitine and its Precursors by Microorganisms and Enzymes

Besides different chemical procedures there are a variety of microbial and enzymatical possibilities which allow the resolution of racemic mixtures of carnitine and its precursors. These procedures are based on the following principles (Fig. 1):

- Enantioselective assimilation of D-carnitine by microorganisms (reaction 1).
- Enantioselective synthesis or hydrolysis of esters of carnitine and its precursors (reactions 2 and 3).
- Enantioselective hydration of racemic mixtures of carnitine precursors (reactions 4, 5 and 6).

Examples for these methods are also listed in Table 1 and will be discussed in the following. Regarding the microbiological procedures most of the enzymes involved are unknown up till now.

2.1 Assimilation of D-Carnitine

The knowledge of D-carnitine metabolism in microorganisms is very limited. In 1977, Kleber et al. [42] described an *Acinetobacter* strain (*Acinetobacter calcoaceticus* 69V) that was able to split the C–N-bond of L- and D-carnitine:

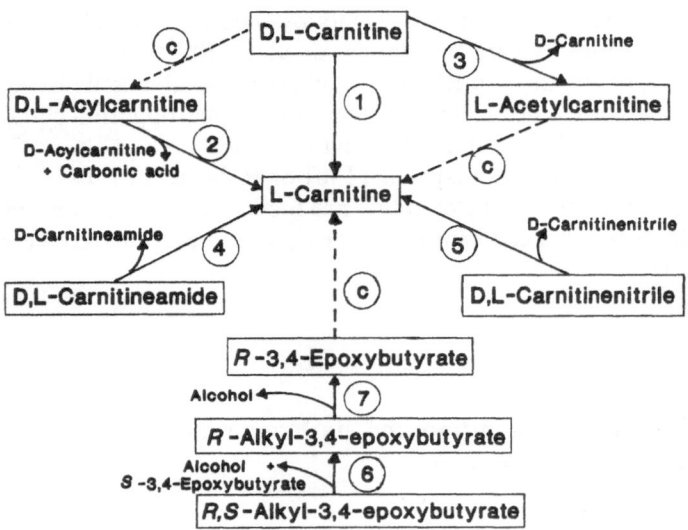

$$\underset{\text{L-Carnitine}}{CH_3-\overset{\overset{\displaystyle CH_3}{|}}{\underset{\underset{\displaystyle CH_3}{|}}{N^+}}-CH_2-\overset{\overset{\displaystyle OH}{|}}{CH}-CH_2-COO^-} \longrightarrow \underset{\text{Trimethylamine}}{CH_3-\overset{\overset{\displaystyle CH_3}{|}}{\underset{\underset{\displaystyle CH_3}{|}}{N}}} + {}^{\cdot}C_4^{\cdot} \qquad (1)$$

Fig. 1. Resolution of D,L-carnitine and racemic precursors by microorganisms and enzymes. Reactions: D-carnitine assimilating microorganisms (*1*); carnitine ester hydrolyzing enzymes and microorganisms (*2*); L-carnitine acetyltransferase (*3*); L-carnitine amidohydrolase (*4*); L-carnitine specific nitrilase (*5*); lipase (*6*); alcalase (*7*) and chemical steps (*c*)

Table 1. Microorganisms which may be applied to the formation of L-carnitine

Reaction type	Starting substrate(s)	Microorganism	Ref.
Enantioselective assimilation of D-carnitine; (involved reactions are unknown)	D,L-carnitine	*Acinetobacter calcoaceticus* ATCC 39647 *A. calcoaceticus* ATCC 39648 *A. lwoffi* ATCC 39769 *A. lwoffi* ATCC 39770	47, 48
Enantioselective ester hydrolysis	D,L-octanoyl-carnitine	*Pseudomonas fluorescens* IMAM *Rhodotorula gracilis* IMAM *Fusarium oxisporum* sp. *lini* IMAM	49
	D,L-acetyl-carnitine	*Escherichia coli* IFO 3301	50

Table 1. (continued)

Reaction type	Starting substrate(s)	Microorganism	Ref.
Enantioselective amide hydrolysis	D,L-carnitine-amide	*Pseudomonas* CA 28-50A	73, 74
Enantioselective nitrile hydrolysis	D,L-carnitine-nitrile	*Corynebacterium* 6N-23	78
Enantioselective reduction	3-dehydro-carnitine	*Rhizobium meliloti*	94
	alkylesters of 4-chloroaceto-acetate or 4-azidoaceto-acetate	*Saccharomyces cerevisiae*	38, 100
Enantioselective hydration	crotonobetaine	*Escherichia coli* 044 K74	139–141
		E. coli K12	139, 140
		E. coli AJ-2590	143
		E. coli AJ-2603	143
		E. coli AJ-2604	143
		E. coli AJ-2621	143
		E. coli AJ-2622	143
		E. coli K12 IFO 3301	143
		E. coli B IFO 13168	143
		Proteus vulgaris	139, 140
		P. vulgaris IFO 3045	144
		P. mirabilis AJ 2772	143
		P mirabilis AJ 12137	143
		Salmonella typhimurium LT_2	139, 140
		S. typhimurium IFO 12529	144
Acyl-CoA synthesis, oxidation (if γ-butyrobetaine is the substrate), enantioselective hydration, acyl-CoA ester hydrolysis	γ-butyro-betaine, crotonobetaine	*Agrobacterium* sp. HK4 *Agrobacterium* sp. HK13	120–122

The assimilation of these quaternary ammonium compounds was accompanied by stoichiometric formation of trimethylamine [43]. The postulated "C_4"-compound is unknown as yet. Resting cells of *A. calcoaceticus* [44] as well as cell-free extracts containing the postulated C–N splitting enzyme [45, 46] could utilize both L- and D-carnitine as substrates at nearly the same rate. According to these findings *A. calcoaceticus* 69V cannot be used to resolve D,L-carnitine.

Sih [47, 48] described a procedure that involves the growth of organisms of the genus *Acinetobacter* (*A. calcoaceticus* ATCC 39647, *A. calcoaceticus* ATCC 39648, *A. lwoffii* ATCC 39769, *A. lwoffii* ATCC 39770) on D,L-carnitine. These strains are characterized by the unique ability to degrade D-carnitine preferentially from the racemate and allow the subsequent recovery of the L-enantiomer:

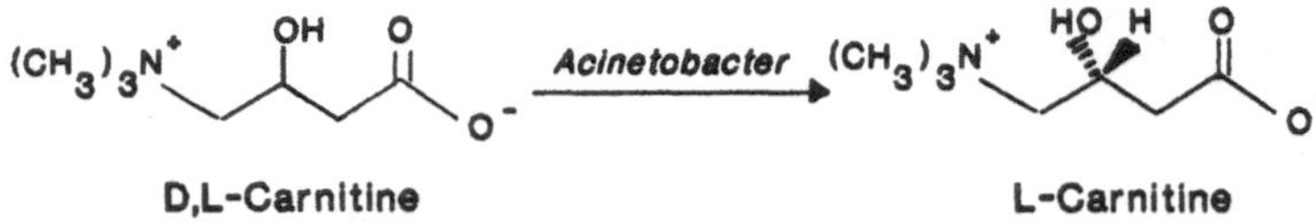

$$D,L\text{-Carnitine} \xrightarrow{\textit{Acinetobacter}} L\text{-Carnitine} \qquad (2)$$

By aerobic incubation of *A. calcoaceticus* ATCC 39647 in the presence of $20\ \mathrm{g\ l^{-1}}$ D,L-carnitine · HCl at 25 °C for 44 h it was possible to produce L-carnitine · HCl with a yield of 38%. The optical purity was estimated to be greater than 96.5% [47]. Information on the enzymatical background of this D-carnitine assimilation are not available up to now.

2.2 Enantioselective Hydrolysis of Esters of Racemic Carnitine and its Precursors

Another possibility of resolving racemic carnitine involves the enantioselective hydrolysis of the corresponding acyl esters of carnitine by microorganisms and enzymes:

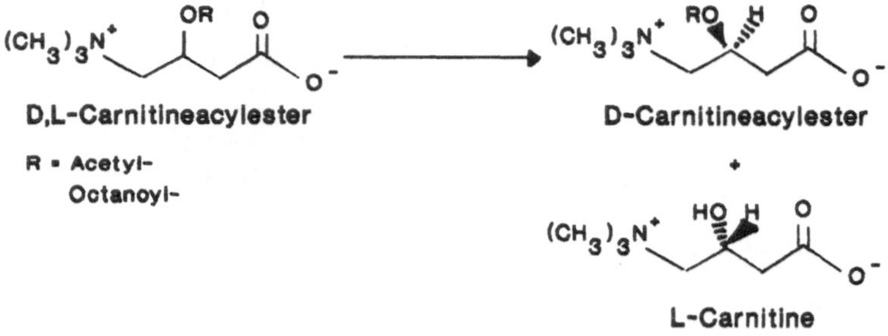

D,L-Carnitineacylester   D-Carnitineacylester

R = Acetyl-
     Octanoyl-

+   (3)

L-Carnitine

Aragozzini et al. [49] found that *Pseudomonas fluorescens* IMAM, *Rhodotorula gracilis* IMAM and *Fusarium oxysporum* sp. *lini* IMAM selectively hydrolyze L-octanoylcarnitine. By incubation of a submerged culture of *F. oxysporum* with D,L-octanoylcarnitine for 24 h L-carnitine was obtained with a yield of 40%. From the specific rotation of the preparation an optical purity of 95% was determined [49]. In a similar way an *Escherichia coli* strain (*E. coli* IFO 3301) was applied to enantioselective hydrolysis of D,L-acetylcarnitine [50].

With respect to the structural similarities between carnitine and choline, Dropsy and Klibanov [51] proposed a method for preparative resolution of racemic carnitine that is based on the application of choline esterases. They demonstrated that acetylcholine esterase [EC 3.1.1.7] from electric eel hydrolyzes the D- but not the L-enantiomer of D,L-acetylcarnitine. On the other hand, only the L-enantiomer of D,L-butyrylcarnitine was hydrolyzed if the racemic mixture was incubated with butyrylcholine esterase [EC 3.1.1.8] from horse serum. For example, the incubation of a 1 M solution of D,L-acetylcarnitine (40 ml) with acetylcholine esterase (1800 units) covalently attached to alumina yielded a complete hydrolysis of the D-enantiomer at 37 °C after 13 h. After separation by ion exchange chromatography and recrystallization L-acetylcarnitine was obtained with a yield of 30% (related to the original racemate) and an optical purity of 88% [51].

Another procedure of L-carnitine synthesis starts from 3,4-epoxybutyric acid esters. Thus, (R, S)-isobutyryl-3,4-epoxy butyric acid was enantioselectively hydrolyzed by a lipase (steapsin from porcine pancreas (Sigma Chemical Co.)) to give the optically active unreacted ester of (R) absolute configuration with an enantiomeric excess greater than 95% [52, 53]. After purification (R)-isobutyryl-3,4-epoxy butyric acid was obtained with a yield of 22% (related to the original racemate). This ester was in a further reaction hydrolyzed by a second enzyme, called alcalase (Novo Industri, Denmark). After heating with trimethylamine and treatment with hydrochloric acid, L-carnitine was obtained. No racemization was observed during the second hydrolysis or the following reactions [52, 53].

2.3 Enantioselective Synthesis of Carnitine Esters

Beside the specific hydrolysis of esters of carnitine or its precursors there is the possibility of enantioselective esterification of carnitine by carnitine acetyltransferase [EC 2.3.1.7] [54]. This enzyme which is well known to occur in several tissues of mammals and birds catalyzes the reversible acetyl transfer between L-carnitine and coenzyme A [55–59]:

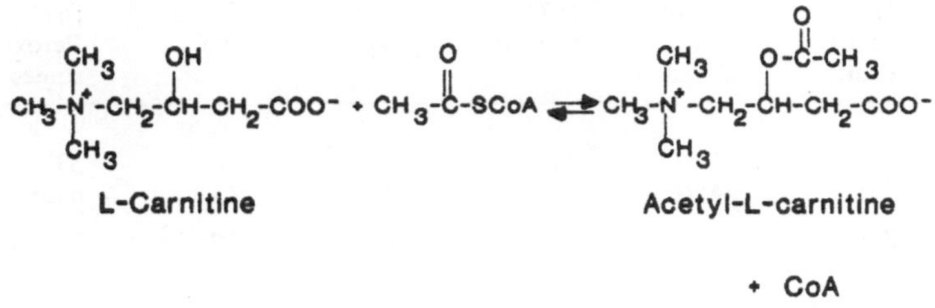

$$CH_3-\overset{\overset{\displaystyle CH_3}{|}}{\underset{\underset{\displaystyle CH_3}{|}}{N^+}}-CH_2-\overset{\overset{\displaystyle OH}{|}}{CH}-CH_2-COO^- \; + \; CH_3-\overset{\overset{\displaystyle O}{||}}{C}-SCoA \; \rightleftharpoons \; CH_3-\overset{\overset{\displaystyle CH_3}{|}}{\underset{\underset{\displaystyle CH_3}{|}}{N^+}}-CH_2-\overset{\overset{\displaystyle O-\overset{\overset{\displaystyle O}{||}}{C}-CH_3}{|}}{CH}-CH_2-COO^- \tag{4}$$

L-Carnitine   Acetyl-L-carnitine

+ CoA

The existence of carnitine acetyltransferase in microorganisms was first demonstrated in the n-alkane-grown yeast *Candida tropicalis* [60]. Ratledge and Gilbert [61] determined carnitine acetyltransferase activity in several non-oleaginous (*Candida utilis, Saccharomyces cerevisiae*) and in various oleaginous yeasts (*Candida curvata, Lipomyces starkeyi, Rhodosporidium toruloides and Trichosporum cutaneum*). Ueda et al. [62] postulated that carnitine acetyltransferase from *C. tropicalis* is involved in a carnitine-dependent shuttle system. This shuttle system may play a role in the transport of acetyl and propionyl residues from peroxisomes (place of β-oxidation of fatty acids in alkane-grown *C. tropicalis*) to mitochondria (place of citric acid cycle) [62].

The carnitine acetyltransferases from *C. tropicalis* [62, 63] and *S. cerevisiae* [64] were purified and characterized (Table 2). Peroxisomal and mitochondrial carnitine acetyltransferases from *C. tropicalis* were separated by ion exchange chromatography on DEAE-Sephacel [62]. Except for their cellular localization and elution pattern there were no great differences in the properties of both enzymes. Analysis by sodium dodecylsulfate polyacrylamide gel electrophoresis of both carnitine acetyltransferases demonstrated that each enzyme preparation consists of two kinds of subunits of different relative molecular masses [62, 65]. The enzyme from rat liver mitochondria is also composed of two distinct subunits [66], but these subunits are smaller as those from the yeast enzymes. Beside peroxisomal and mitochondrial forms of carnitine acetyltransferase Kozulic et al. [67] describe a soluble form of this enzyme in *C. tropicalis*. The most apparent difference between these forms exists in their quaternary structure. The cytosolic enzyme seems to be an octamer composed of subunits equal in size whereas both peroxisomal and mitochondrial carnitine acetyltransferases have two types of subunits differing in size [67]. However, the relative molecular mass of the soluble carnitine acetyltransferase subunit (64 kDa) is the same as the relative molecular mass of the larger subunit present in the mitochondrial and peroxisomal enzymes [67]. Kispal et al. [64] assume that the high relative molecular mass of native carnitine acetyltransferase from *S. cerevisiae* (400 kDa) may probably be due to an artefact because this enzyme has the tendency to polymerize and to precipitate at high protein concentrations. The K_m values of

Table 2. Comparison of carnitine acetyltransferase from different sources

Properties	*Candida tropicalis* [62, 65] Peroxi- somes	*Candida tropicalis* [62, 65] Mito- chondria	*Candida tropicalis* [67] Cytosol	*Saccharo- myces cerevisiae* [64][a]	Pigeon breast [68, 69, 70][a]	Rat liver [69] Peroxi- somes	Rat liver [66, 68] Mito- chondria
M_r (kDa):							
I[b]	420	420	540	400	51	59	62
II[c]	64/57	64/52	64	65	75	n.d.[d]	36/27
pH- Optimum	8.0	8.0	n.d.	7.5–8.0	7.2–7.8	7.2–7.8	8.0
K_m values (mM):							
L-carnitine	0.72	0.62	n.d.	0.17	0.12	0.14	0.72[e]
L-acetyl- carnitine	0.42	0.64	n.d.	n.d.	0.35	n.d.	0.28
CoA	0.30	0.26	n.d.	n.d.	0.04	n.d.	0.03
acetyl-CoA	0.04	0.04	n.d.	0.02	0.03	0.07	0.03

[a] Regarding the parameters listed in the table the authors did not distinguish between enzymes from mitochondria and cytosol
[b] Determination by gel filtration
[c] Determination by sodium dodecylsulfate polyacrylamide gel electrophoresis
[d] Not determined
[e] The substrate was D,L-carnitine

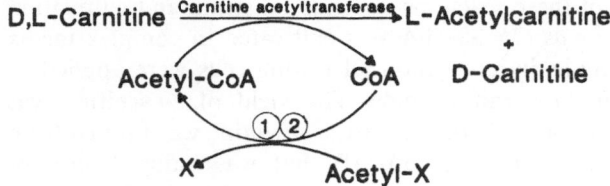

Fig. 2. Resolution of racemic carnitine by enantioselective acetylation. Regeneration of acetyl-CoA: enzymatical acetyltransfer by phosphotransacetylase, $x = PO_4$ (*1*) [54]; chemical reaction, $x = $ thiocholine (*2*) [72]

carnitine acetyltransferases from different sources have the same order of magnitude (Table 2). Compared with the enzymes from mammalian tissues both enzymes are specific only to acetyl and propionyl groups and showed no activity with medium-chain and long-chain acyl groups [62].

Carnitine acetyltransferase was used for the enantioselective acetylation of L-carnitine [54]. The donor of the acetyl residue – acetyl CoA – may be recycled by a system composed of phosphotransacetylase [EC 2.3.1.8], acetylphosphate and CoA (Fig. 2). In this way, 60% of the L-carnitine of a racemic carnitine probe were converted into L-acetylcarnitine [54]. After separation of the L-isomer from carnitine no detectable amounts of D-carnitine were shown by ^{1}H-NMR analysis using the chiral shift reagent tris[3-((tri-fluoromethyl)hydroxy-methylene)-d-camphorato]eurobium (III) [71]. Another way for acetyl-CoA recycling involves chemical acylation of CoA by means of S-acetyl thiocholine iodide [72].

2.4 Enantioselective Hydrolysis of Carnitineamide and Carnitinenitrile

The patent literature offers two further ways to synthesize optical pure forms of carnitine starting from racemic precursors. One method involves the enzymatic hydrolysis of carnitineamide. Nakayama et al. [73, 74] describe different *Pseudomonas* species which hydrolyze L- or D-carnitineamide with high enantio-selectivity:

D,L-Carnitineamide

D-Carnitineamide

+ (5)

L-Carnitine

The enantioselectivity of these reactions depends on strain and cultivation conditions. Thus, *Pseudomonas* CA 28-50A was cultivated in complex media containing L-carnitineamide as inducer. Induced resting cells were applied to L-carnitine synthesis from D,L-carnitineamide. The yield of L-carnitine was 81.6% with an optical purity of 99%. In contrast, D-carnitine was formed from D,L-carnitineamide by *Pseudomonas* CA 30-11B that was induced with D-carnitineamide. The corresponding enzymes, termed carnitineamide hydrolases, were purified and characterized [75]. The hydrolyzed enantiomer may be separated from the surviving carnitineamide by ion exchange chromatography [76, 77].

Similar to these findings L-carnitine may be produced by hydrolysis of D,L-carnitinenitrile with a nitrilase from *Corynebacterium* species [78]. The enzyme seems to be an inducible one accepting D- and L-carnitine as inducers. The obtained yield of L-carnitine was approximately 78% [78].

3 Enantioselective Synthesis of L-Carnitine from Achiral Precursors

L-Carnitine production via resolution of racemic carnitine allows only the utilization of a maximum of 50% of the original substrates. One alternative to overcome this disadvantage consists of the enantioselective L-carnitine synthesis from achiral precursors. With respect to the present knowledge on carnitine metabolism in microorganisms there are three general possibilities for the usage of biological systems for this purpose (Fig. 3):

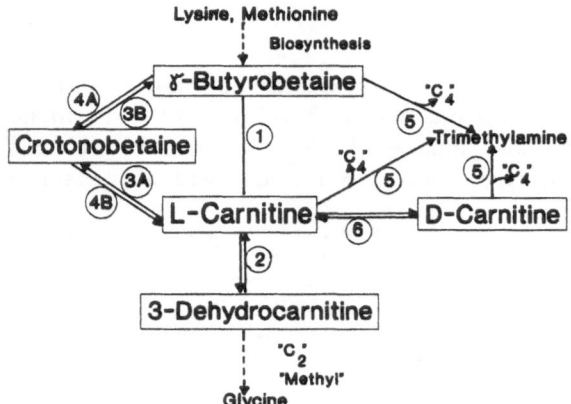

Fig. 3. Microbial metabolism of L-carnitine. Reactions are catalyzed by the following enzymes (-systems): γ-butyrobetaine hydroxylase (*1*); L-carnitine dehydrogenase (*2*); carnitine dehydratase (*3A*); "crotonobetaine reductase" (*3B*); "γ-butyrobetainyl-CoA dehydrogenase" (*4A*); "carnitinyl-CoA hydrolase" (*4B*) "carnitine CN-lyase" (*5*); "carnitine recemase" (*6*); postulated enzymes in quotation marks

1. Usage of enzymes which are involved in L-carnitine biosynthesis from lysine and methionine (reaction 1).
2. Application of microorganisms and enzymes which are involved in the assimilation of achiral quaternary ammonium compounds thereby L-carnitine is an intermediate of the degradation pathway. This pathway must be blocked after the step of L-carnitine formation if intact cells are applied for L-carnitine synthesis (reaction 4).
3. Return of reactions which are involved in L-carnitine degradation under physiological conditions (reactions 2 and 3A).

Another group of methods of L-carnitine synthesis from achiral precursors involves microbial reduction of acetoacetic acid ester derivatives followed by different chemical reactions.

Examples for all these procedures are listed in Table 1.

3.1 Enantioselective Reduction of Keto Acids

3.1.1 Reduction of 3-Dehydrocarnitine by L-Carnitine Dehydrogenase

Different *Pseudomonas* species are able to grow aerobically on L-carnitine as sole source of carbon and nitrogen [79]. The first catabolic step is the oxidation of the β-hydroxy group of L-carnitine with concomitant formation of 3-dehydrocarnitine:

L-Carnitine   3-Dehydrocarnitine

$$(6)$$

This reaction is catalyzed by L-carnitine dehydrogenase [EC 1.1.1.108] [80]. L-Carnitine and 3-dehydrocarnitine are inducers of this enzyme in *Pseudomonas* species [81, 82]. L-Carnitine dehydrogenase was first isolated from *P. aeruginosa* A 7244 [80]. This enzyme has also been purified from *P. putida* [83], *Xanthomonas translucens* [84] and recently from *Alcaligenes* [85]. The L-carnitine dehydrogenase gene from *X. translucens* was cloned and expressed in *Escherichia coli* [86]. The properties of the investigated enzymes of different microorganisms are summarized in Table 3. All of the isolated enzymes are very specific for L-carnitine and NAD$^+$. The L-carnitine dehydrogenases from *P. putida* and from *X. translucens* are dimeric with relative molecular masses of 62 and 74 kDa, respectively, whereas L-carnitine dehydrogenase from *Alcaligenes* has a relative molecular mass of 51 + 6 kDa. The two *Pseudomonas* enzymes and the L-carnitine dehydrogenase of *X. translucens* have very similar kinetic properties and the Michaelis constants are very close. The equilibrium constant of the oxidation of L-carnitine was determined with $K_{eq} = 1.3 \times 10^{-11}$ [80].

Table 3. Properties of L-carnitine dehydrogenase isolated from different microorganisms

Properties	Pseudomonas aeruginosa [80, 87]	Pseudomonas putida [83]	Xanthomonas translucens [84]	Alcaligenes [85]
Substrate specifity	L-carnitine NAD$^+$	L-carnitine NAD$^+$ weak activity with 4-amino-3-hydroxybutyrate	L-carnitine NAD$^+$	L-carnitine NAD$^+$
M$_r$ (kDa)	n.d.[a]	62 (2 subunits)	74 (2 subunits)	51
IP	n.d.	4.7	n.d.	5.3 + 0.6
Temperature optimum	35 °C	30 °C	n.d.	50 °C
pH-Optimum: oxidation reduction	9.0 7.0	9.0 7.0	9.5 6.5	n.d. n.d.
K$_m$ values (mM): L-carnitine NAD$^+$ 3-dehydro-carnitine NADH$^+$	8.90 0.16 1.54 0.02	6.25 0.20 1.54 0.07	10.00 0.25 1.71 0.04	n.d. n.d. n.d. n.d.

[a] Not determined

Position of equilibrium and stability of L-carnitine dehydrogenase favor the application of this enzyme for synthetic purposes. On the other hand the substrate 3-dehydrocarnitine is relatively unstable.

In 1980, Vandecasteele and Lemal [88] published a method for L-carnitine manufacture with 3-dehydrocarnitine as precursor. For this purpose L-carnitine dehydrogenase from induced cells of *P. putida* CIP 52 191 was used for synthesis. Because of fast decarboxylation of 3-dehydrocarnitine this compound was continuously introduced into the reaction mixture as highly acidic solution as a rate limiting constituent.

There are different ways to regenerate the necessary coenzyme NADH [89]. The most successful methods involve enzymatical systems. One example is the use of glucose dehydrogenase [EC 1.1.1.47] [88] (Fig. 4). This enzyme can be produced in a cheap manner by *Bacillus megaterium* ATCC 39118 [90]. Applying this NADH regeneration system L-carnitine was produced from 3-dehydrocarnitine with a yield of 95% [88].

Another favored regeneration system involves the use of formate dehydrogenase [EC 1.2.1.2] [88, 91] which catalyzes the following reaction:

$$HCOOH + NAD^+ \longrightarrow CO_2 + NADH + H^+ \tag{7}$$

The advantage of this reaction is the absence of reaction products which have to be separated. Souppe et al. [92] describe that manufacturing of

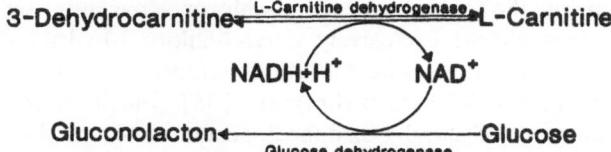

Fig. 4. L-Carnitine synthesis from 3-dehydrocarnitine with regeneration of NADH by glucose dehydrogenase [88]

L-carnitine using formate dehydrogenase as regenerating system is improved if the ionic strength is raised during the reaction up to more than 0.5 M. With this method the amount of L-carnitine dehydrogenase used can be reduced and the stability of the enzyme is increased. Thus, a 98% reduction and output of 1.67 g L-carnitine $h^{-1} l^{-1}$ was obtained using a ratio of 868 units L-carnitine dehydrogenase to 1 mol 3-dehydrocarnitine, an initial ionic strength of 0.20 M and a final ionic strength of 1.13 M [92].

Also a NAD^+-polyethylene glycol conjugate was successfully used in L-carnitine synthesis [93].

Donishi and Yokozeki [94] produced L-carnitine from 3-dehydrocarnitine with whole cells of *Rhizobium meliloti*. The yield obtained was 65%.

It should be mentioned that L-carnitine dehydrogenase was applied for analytical purposes. A spectrophotometric test for the quantitative estimation of L-carnitine was elaborated [95]. Also, an amperometric bienzyme electrode for L-carnitine determination, using L-carnitine dehydrogenase and diaphorase, was developed [96].

3.1.2 Reduction of Acetoacetate Derivatives by Microorganisms

Generally, oxidoreductases are suited for synthetic pathways because they accept a broad range of compounds as substrates, and many of them are highly enantioselective [97–99]. These properties were used to develop chemomicrobiological methods of L-carnitine production [38, 100, 101]. The principle of these methods is the reduction of acetoacetic acid esters into the corresponding chiral alcohol, precursor of L-carnitine, catalyzed by yeasts:

Alkylchloroacetoacetate

R-Alkylchloroacetoacetate

R = octyl-
  ethyl-

(8)

L-Carnitine

Interestingly, reduction of the octyl ester of 4-chloroacetoacetate by *Saccharomyces cerevisiae* gives almost exclusively octyl-4-chloro-3-hydroxy-butyrate of (*R*) absolute configuration, whereas ethyl-4-chloroacetoacetate is reduced preferentially to (*S*)-ethyl-3-hydroxybutyrate [38]. Furthermore, *S. cerevisiae* mediated reduction of ethyl-4-azidoacetoacetate and ethyl-4-bromoacetoacetate affords (3*R*) and (3*S*) alcohols, respectively, in high optical purity [100]. Thus, *S. cerevisiae* contains different enzymes acting in accordance to substrate specifity and kinetical properties with opposite stereochemistry [102].

For example, octyl-4-chloroacetoacetate was added to an actively growing culture of *Candida kefyr* and incubated for 24 h. The conversion yield into the corresponding enantiomeric alcohol was 79% [101]. Bare et al. [103] used *S. cerevisiae* and detected 90–95% conversion of the same substrate thereby the enantiomeric purity was almost 100%. Immobilized cells in alginate beads were less efficient [103]. The limiting factor of the reduction process seems to be the turnover of involved cofactors [103]. (*R*)-Octyl-4-chloro-3-hydroxybutyrate can be transformed into L-carnitine by heating with trimethylamine and hydrolysis with hydrochloric acid [38].

3.2 L-*Carnitine Synthesis from* γ-*Butyrobetaine*

The last step of the L-carnitine biosynthetic pathway is the hydroxylation of γ-butyrobetaine to L-carnitine [1]:

$$CH_3-\overset{\overset{\displaystyle CH_3}{|}}{\underset{\underset{\displaystyle CH_3}{|}}{N^+}}-CH_2-CH_2-CH_2-COO^- + O_2 + {}^-OOC-CH_2-CH_2-\overset{\overset{\displaystyle O}{\|}}{C}-COO^-$$

γ-Butyrobetaine   α-Oxoglutarate

$$\longrightarrow CH_3-\overset{\overset{\displaystyle CH_3}{|}}{\underset{\underset{\displaystyle CH_3}{|}}{N^+}}-CH_2-\overset{\overset{\displaystyle OH}{|}}{CH}-CH_2-COO^- + CO_2 + {}^-OOC-CH_2-CH_2-COO^-$$

L-Carnitine   Succinate

(9)

This reaction is catalyzed by γ-butyrobetaine hydroxylase [EC 1.14.11.1] [104]. This enzyme has an absolute requirement for ferrous ions, and belongs to a unique class of non-heme ferrous ion dioxygenases in which the hydroxylation of substrate is linked to the oxidative decarboxylation of α-oxoglutarate [105, 106]. γ-Butyrobetaine hydroxylase has been isolated in homogeneous form from *Pseudomonas* sp. AK1 [107] and calf liver [108]. *Pseudomonas* sp. AK1 is able to grow on γ-butyrobetaine as sole carbon source thereby L-carnitine is an intermediate in the degradation pathway [109].

A reductant such as ascorbate as well as catalase are required for maximal activity of both liver [110] and bacterial enzymes [111]. Rat liver γ-butyrobetaine hydroxylase is more effectively stimulated in the presence of glutathione peroxidase plus reduced glutathione [112]. Furthermore, potassium causes an augmentation of L-carnitine synthesis [113].

The properties of γ-butyrobetaine hydroxylase from different sources are similar (Table 4). The relative molecular masses of both bacterial and liver enzymes are comparable. While the bacterial enzyme consists of two polypeptides of relative molecular masses of 37 and 39 kDa, respectively, the mammalian enzyme appears to be a dimer of subunits of equal size. The Michaelis constant for γ-butyrobetaine is similar to that of γ-butyrobetaine hydroxylase from *Pseudomonas* sp. AK1 and calf liver. Both the bacterial and calf liver γ-butyrobetaine hydroxylase are inhibited by certain metal ions and sulfhydryl reagents. But they show no immunological cross-reactivity [107]. Lindstedt and Nordin [115] identified multiple isoenzymes of γ-butyrobetaine hydroxylase in human kidney and liver. These forms were separated by chromatofocusing (Table 4).

Table 4. Properties of γ-butyrobetaine hydroxylase isolated from different sources

Properties	*Pseudomonas* sp. AK 1 [106, 113]	Calf liver [107]
M_r (kDa)	82 (2 subunits; 37/39)	80 (2 identical subunits)
IP	5.1	5.6; 5.7; 5.8 [115]
Apparent K_m values (mM):		
γ-butyrobetaine	2.40	0.51
α-oxoglutarate	0.45	0.82
Fe^{2+}	0.06	0.01
Activation constants:		
ascorbate	n.d.[a]	5.10 mM
catalase	n.d.	0.92 µM
Inhibitors	-divalent cations (Cu^{2+}, Ni^{2+}, Hg^{2+} CO^{2+}, Zn^{2+}, Cd^{2+}) -sulfhydryl reagents (*p*-mercuriphenyl- sulfonate, *p*-mercuribenzoate)	-divalent cations (Cu^{2+}, Ni^{2+}, Hg^{2+} CO^{2+}, Zn^{2+}) -sulfhydryl reagents (α,α'-bipyridyl, metal chelators)
Immunological properties		no cross reactivity with purified γ-butyrobetaine hydroxylase from *Pseudomonas*

[a] Not determined

γ-Butyrobetaine hydroxylase has been the subject of several mechanistic studies in recent years. It has been shown that the enzyme is indeed a dioxygenase by means of $^{18}O_2$ labeling [116] and the hydroxylation has been found to proceed with retention of configuration C-3 [117]. Furthermore, there is an uncoupling of α-oxoglutarate decarboxylation from γ-butyrobetaine hydroxylase [118].

The enantioselective hydroxylation of γ-butyrobetaine has also been used for L-carnitine production. Cavazza [119] used the γ-butyrobetaine hydroxylase isolated from spores of *Neurospora crassa*. Cell-free extract was incubated with γ-butyrobetaine, α-oxoglutarate, Fe^{2+}, ascorbate and catalase to give L-carnitine in a yield of 80%. The storage stability of the used spores is an advantage of this procedure.

An excellent biotechnological procedure for L-carnitine production from crotonobetaine and γ-butyrobetaine was developed by Kulla et al. [120–122]. They isolated a microorganism from soil that was able to grow on L-carnitine, γ-butyrobetaine, crotonobetaine and glycinebetaine but not on D-carnitine as sole source of carbon and nitrogen under aerobic conditions. Taxonomically the strain was situated between *Agrobacterium* and *Rhizobium* [122]. It assimilated L-carnitine similar as shown for different *Pseudomonas* species including the oxidation of L-carnitine by a specific dehydrogenase [79–82]. γ-Butyrobetaine was not metabolized by a specific hydroxylase as demonstrated for *Pseudomonas* sp. AK 1 [107, 109]. Rather γ-butyrobetaine was degraded via a reaction sequence that may be compared with the β-oxidation of fatty acids (Fig. 5). Thus, γ-butyrobetainyl-CoA is synthesized in a first step followed by dehydrogenation whereby forming crotonobetainyl-CoA. Subsequently L-carnitinyl-CoA was synthesized by a hydrolyase, and the thioester bond was cleaved by a thioesterase [122]. The corresponding enzymes were enriched, but purification was hampered because of their instability. γ-Butyrobetainyl-CoA synthetase and the hydrolyase seem to form a complex which is easily separated from γ-butyrobetainyl-CoA dehydrogenase. Furthermore, the thioesterase seems to be integrated into the complex [122].

A mutant of the isolated strain HK4 was produced that did not possess L-carnitine dehydrogenase activity. This mutant (termed HK13) was applied to L-carnitine synthesis from γ-butyrobetaine as well as crotonobetaine. The yield of L-carnitine was nearly 100%. The product was enantiomerically pure [120–122]. The best results were obtained with glycine betaine as the growth substrate.

Agrobacterium sp. strain HK4 possesses a specific and energy consuming transport system for γ-butyrobetaine that is induced by L-carnitine, crotonobetaine and γ-butyrobetaine [123]. This transport system depends on a periplasmic binding protein [124–127].

Based on these findings a procedure was developed which allows L-carnitine production from γ-butyrobetaine in a continuous manner [122]. A high efficiency of the biotransformation is realized by a cell recycling system which provides high biomass concentrations (Fig. 6). Cells in this bioreactor show high metabolic activity at low growth rates. Starting from γ-butyrobetaine the obtained yield of L-carnitine is approximately 92% [121]. Compared with this

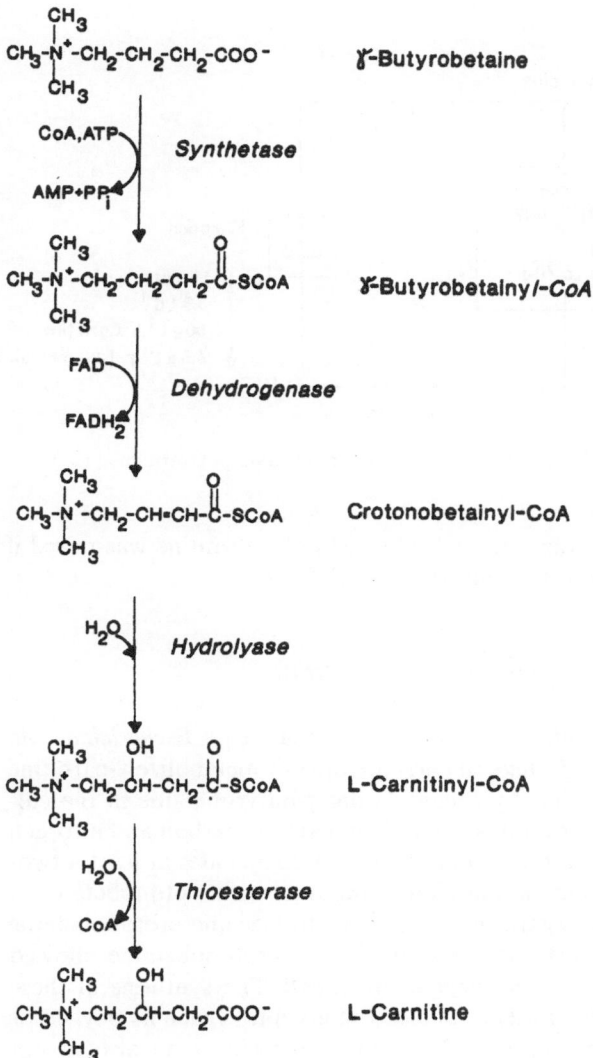

Fig. 5. L-Carnitine synthesis from γ-butyrobetaine and crotonobetaine by an *Agrobacterium*-like strain [122]

procedure fed-batch operation provides higher product yields (approx. 99.5%) whereas the volume productivity is significantly lower [122].

Furthermore, the pathway of L-carnitine biosynthesis was used for the production of this quaternary ammonium compound by growing microorganisms. For example *Rhizopus oligosporus* IFO 8631 [128], *Saccharomyces cerevisiae* [129], *Monascus anka* IFO 4478 [130] are cultivated in a complex medium to grow and synthesize L-carnitine. With the latter strain a yield of

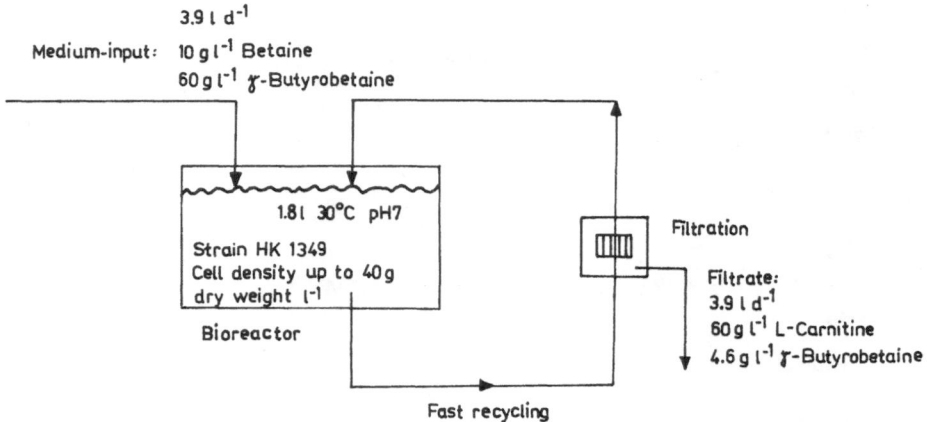

Fig. 6. Laboratory scale continuous recycling culture for the production of L-carnitine [122]

164 μg L-carnitine g^{-1} cells was achieved. The yield of L-carnitine was raised if γ-butyrobetaine was added to the culture medium [129].

3.3 L-*Carnitine Synthesis from Crotonobetaine*

Various members of the family of Enterobacteriaceae, e.g., *Escherichia coli*, *Salmonella typhimurium* and *Proteus vulgaris* are able to metabolize L-carnitine under anaerobic conditions thereby accumulating γ-butyrobetaine in the culture medium [131–135]. These strains fail to assimilate the carbon and nitrogen skeleton of carnitine. The metabolization of L-carnitine includes at least a two-step reduction of this quaternary ammonium compound with crotonobetaine as intermediate (Fig. 7). Two Enzymes, carnitine dehydratase and crotonobetaine reductase, catalyze these reactions. L-Carnitine and crotonobetaine showed themselves to be inducers of both enzymes in *E. coli*. The synthesis of these enzymes was repressed in the presence of electron acceptors such as oxygen or nitrate as well as by glucose [136, 137]. Fumarate, which may also accept electrons under anaerobic conditions, did not repress the enzymes but inhibited the reduction of crotonobetaine to γ-butyrobetaine. In the absence of other electron acceptors (oxygen, nitrate, trimethylamine oxide, fumarate) the addition of L-carnitine or crotonobetaine to the culture medium caused a significant stimulation of the cell growth [138].

.Carnitine dehydratase was isolated for the first time from *E. coli* 044 K74 and characterized [136]. The enzyme catalyzes specifically the reversible dehydration of L-carnitine thereby forming crotonobetaine. The equilibrium constant of crotonobetaine hydration was determined to be approx. 1.5. The relative molecular mass of the enzyme was estimated to be 85 kDa. It seems to be composed of two identical subunits with a relative molecular mass of 45 kDa.

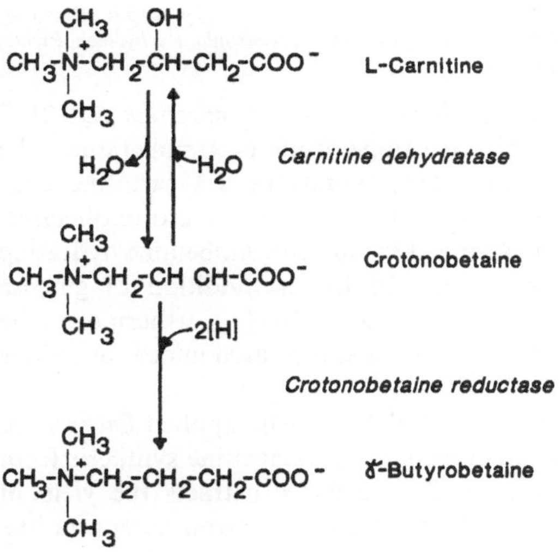

CH₃ OH
|⁺ |
CH₃-N-CH₂-CH-CH₂-COO⁻   L-Carnitine
|
CH₃

H₂O ↗ ↖ H₂O   *Carnitine dehydratase*

CH₃
|⁺
CH₃-N-CH₂-CH CH-COO⁻   Crotonobetaine
|
CH₃   ⟵-2[H]

Crotonobetaine reductase

CH₃
|⁺
CH₃-N-CH₂-CH₂-CH₂-COO⁻   γ-Butyrobetaine
|
CH₃

Fig. 7. Metabolization of L-carnitine in Enterobacteriaceae

The optimum reaction conditions for enzymatic crotonobetaine hydration were found at a pH of 7.8 in potassium phosphate buffer and at a temperature between 37 and 42 °C. γ-Butyrobetaine, D-carnitine and choline proved as competitive inhibitors of crotonobetaine hydration [136].

Whole resting and immobilized cells of different enteric bacteria, particularly of the genera *Escherichia*, *Proteus* and *Salmonella*, were applied to enantioselective L-carnitine synthesis from crotonobetaine [139–142]. Induced resting cells of *E. coli* 044 K74 produced L-carnitine from crotonobetaine with a maximum yield of approx. 60%. The enantiomerical purity was approx. 98%. Electron acceptors like nitrate or fumarate were added as inhibitors of γ-butyrobetaine formation. In a continuous method using cells immobilized by gel entrapment L-carnitine was produced from crotonobetaine up to a yield of 40% [141].

In principle, there is the possibility of L-carnitine synthesis from crotonobetaine by means of the isolated and immobilized carnitine dehydratase. However, this method proved as irrelevant for large scale L-carnitine production since the isolated enzyme is unstable and must be reactivated by an – up to now – unknown cofactor [142].

Yokozeki et al. [143] tested 706 strains of microorganisms on their ability to synthesize L-carnitine from crotonobetaine. The authors demonstrated that 2 strains of actinomycetes, 126 strains of bacteria and 20 strains of yeasts convert crotonobetaine into L-carnitine if the microorganisms were precultivated in the presence of crotonobetaine and subsequently suspended in a phosphate buffer containing crotonobetaine.

These L-carnitine producers belong to the following genera [143]: *Norcardia, Streptomyces, Achromobacter, Alcaligenes, Agrobacterium, Aeromonas, Arthrobacter, Acinetobacter, Bacillus, Brevibacterium, Cellulomonas, Citrobacter, Corynebacterium, Erwinia, Escherichia, Enterobacter, Flavobacterium, Hafnia, Kluyvera, Kurthia, Klebsiella, Micrococcus, Microbacterium, Pseudomonas,*

Proteus, Protaminobacter, Serratia, Sarcina, Vibrio, Xanthomonas, Candida, Cryptococcus, Debaryomyces, Geotrichum, Hansenula, Kloeckera, Kluyveromyces, Lipomyces, Nadsonia, Pachysolen, Pichia, Rhodotorula, Saccharomyces, Trigonopsis and Torulopsis.

During the screening, *Proteus mirabilis* J-2772 and *P. mirabilis* AJ 12137 proved to be the best producers of L-carnitine from crotonobetaine. The maximum yield was about 50% which corresponds to a L-carnitine concentration of $40 \, g l^{-1}$ starting from a $62.5 \, g l^{-1}$ solution of crotonobetaine. Similar to that shown for the *E. coli* enzyme [136] the crotonobetaine hydrating enzyme (carnitine dehydratase?) was repressed by the addition of glucose suggesting a regulation by catabolite repression [143]. Furthermore, the crotonobetaine hydrating activity of *P. mirabilis* was increased under conditions of oxygen limitation.

Similar results were obtained by Fukui et al. [144] who applied *Escherichia coli, Salmonella typhimurium* and *Proteus vulgaris* to L-carnitine synthesis from crotonobetaine. Using resting cells as well as cell-free extracts the yield of L-carnitine synthesis was increased by the addition of electron acceptors like nitrate or trimethylamine oxide. Under optimum incubation conditions up to about 50% of a 0.1 M crotonobetaine solution were converted into L-carnitine using cell-free protein extracts of an *E. coli* K12 strain [144]. Furthermore, Kawamura et al. [145–149] describe the synthesis of L-carnitine from crotonobetaine by application of a great variety of strains belonging to different genera of fungi, yeasts and bacteria.

As described above (cf. Sect. 3.2) crotonobetaine may also be converted into L-carnitine by an *Agrobacterium*-like species [120–122].

4 Concluding Remarks

Regarding the physiological importance of L-carnitine and its application in medicine, it is not surprising that a great number of different procedures of L-carnitine production have been proposed. The diversity of pathways which are involved in the metabolism of carnitine and related compounds provides a variety of possibilities to obtain this quaternary ammonium compound in optically pure form.

In general, enzyme catalysis offers different advantages compared with chemical procedures. Enzymes are characterized by high substrate specifity, enantioselectivity and operation at ambient temperature [150]. Thus, enantioselective action of microorganisms and enzymes allows the resolution of racemic carnitine and derivatives (e.g. D,L-acetylcarnitine [51], D,L-carnitineamide [73, 74]). The final accumulation of D-carnitine (or related compounds) may be prevented by the method of Sih [47, 48] which involves enantioselective assimilation of D-carnitine by bacteria. But also this method has the disadvantage that only maximum 50% of the original substrates may be converted into L-carnitine. This problem may be overcome by L-carnitine synthesis from achiral precursors like γ-butyrobetaine, crotonobetaine, 3-dehydrocarnitine and derivatives of acetoacetate. The latter methods provide L-carnitine preparations of very high enantiomerical purity (nearly 100%). High yields of L-carnitine (more than 95%) were obtained, e.g. by L-carnitine synthesis from γ-butyrobetaine using an *Agrobacterium*-like species [120–122] as well as from 3-dehydrocarnitine using L-carnitine dehydrogenase [88, 91]. L-Carnitine synthesis from crotonobetaine is interesting with regard to D-carnitine recycling since D-carnitine can easily be dehydrated to crotonobetaine in chemical manner.

In spite of the advantages mentioned above the application of microorganisms and enzymes for L-carnitine production on an industrial scale is very restricted. Procedures which involve the usage of isolated enzymes are often very complex. Thus, the action of oxidoreductases (e.g. carnitine dehydrogenase [88, 91], γ-butyrobetaine hydroxylase [119]) or transferases (e.g. carnitine acetyltransferase [49]) requires the addition of coenzymes, cosubstrates or other biochemicals. Furthermore, since coenzymes are expensive, appropriate recycling systems have to be developed. With respect to these problems it seems to be easier to use esterases (e.g. acetylcholine esterase which is applied to the resolution of D,L-carnitine [51]) and other hydrolytic enzymes whose activity does not depend on coenzymes and cosubstrates. Other critical points of microbiological and enzymatical procedures concern problems with the long-term stability of biocatalysts as well as the possibility of substrate or product inhibition. Inhibitory effects of the substrate or product on the enzyme activity may be avoided by operation at low substrate concentrations, but the subsequent isolation of the product from diluted solutions often makes the corresponding method very expensive.

Although by-product formation and problems of the transport of substrate and product are often inherent disadvantages of microbiological procedures, whole cells are currently used preferably for L-carnitine formation compared with enzymes. By-product formation may be limited by construction of genetically altered strains. The example of L-carnitine production from γ-butyrobetaine by a mutant of an *Agrobacterium*-like strain [120–122] shows that classical mutagenesis may provide very productive bacteria. Furthermore, whole cells offer the possibility of coenzyme recycling during the assimilation of cheap growth substrates. Under optimal process conditions stable activities of substrate transformation may be established for several weeks allowing a continuous operation.

Besides finding of new enzymatical reactions relevant to L-carnitine synthesis, molecular biology is opening up new possibilities in enantioselective synthesis. This may concern, e.g. the construction of biocatalysts of long-term stability and resistance to high concentrations of substrate and product.

5 References

1. Bremer J (1983) Physiol Rev 63: 1420
2. Gulewitsch W, Krimberg R (1905) Hoppe-Seyler's Z Physiol Chem 45: 326
3. Kutscher FZ (1905) Untersuch Nahr Genussm 10: 528
4. Tomita M, Sendju Y (1927) Hoppe-Seyler's Z Physiol Chem 169: 263
5. Friedman S, Fraenkel GS (1972) Carnitine. In: Sebrell WH, Harris RS (eds) The vitamins, 2nd edn. Academic Press, New York, vol 5, p 329
6. Fritz IB (1963) Adv Lipid Res 1: 285
7. Bieber LL (1988) Ann Rev Biochem 57: 261
8. Borum PR (1983) Ann Rev Nutr 3: 233
9. Scholte HR, de Jonge PC (1987) Metabolism, function and transport of carnitine in health and disease. In: Gitzelmann R, Baerlocher K, Steinman B (eds) Carnitin in der Medizin. Schattauer, Stuttgart, p 21
10. Tanphaichitr V, Broquist HP (1973) J Biol Chem 248: 2176
11. Wolf G, Berger CRA (1961) Arch Biochem Biophys 92: 360
12. Rebouche CJ, Broquist HP (1976) J Bacteriol 126: 1207
13. Borum PR, Broquist HP (1977) J Biol Chem 252: 5651
14. Paik WK, Kim S (1975) Adv Enzymol Relat Areas Mol Biol 42: 227
15. Henderson LM, Nelson PJ, Henderson L (1982) Fed Proc 41: 2843
16. Rebouche CJ, Engel AG (1980) Biochim Biophys Acta 630: 22
17. Bremer J (1962) Biochim Biophys Acta 57: 327
18. Lindstedt G, Lindstedt S (1970) J Biol Chem 245: 4178
19. Rebouche CJ, Engel AG (1980) J Biol Chem 255: 8700
20. Engel AG, Angelini C (1973) Science 179: 899
21. Carroll EC, Carter AL, Perlman S (1987) J Nutr 117: 1501
22. Gilbert EF (1985) Pathology 17: 161
23. Rebouche CJ, Paulson DJ (1986) Ann Rev Nutr 6: 41
24. Hulsmann WC (1991) Cardiovascular Drugs and Therapy 5: 7
25. Meier PJ (1987) D-Carnitin, harmlos? In: Gitzelmann R, Baerlocher K, Steinman B (eds) Carnitin in der Medizin. Schattauer, Stuttgart p 101
26. Emaus KR, Bieber LL (1983) J Biol Chem 258: 13160
27. Kleber HP, Claus R (1982) DD 204378
28. Kleber HP, Claus R (1981) DD 211039
29. Kleber HP, Claus R, Seim H, Strack E (1981) DD 204105
30. Typelt H, Claus R, Nitzsche K (1991) J Biotechnol 18: 173
31. Strack E, Lorenz I (1960) Hoppe-Seyler's Z Physiol Chem 318: 129
32. Strack E, Müller DM (1972) Hoppe-Seyler's Z Physiol Chem 353: 618
33. Cavazza C (1980) Germ Offen DE 2927672
34. Comber RN, Brouillette WJ (1987) J Biol Chem 52: 2311
35. Voeffray R, Perlberger JC, Tenud L, Gosteli J (1987) Helv Chim Acta 70: 2058
36. Bock K, Lundt I, Pederson C (1983) Acta Chem Scand, Ser B 37: 341
37. Fiorini M, Valentini C (1982) Eur Patent Appl EP 60595
38. Zhou B, Gopalan AS, Van Middlesworth F, Shieh WR, Sih CJ (1983) J Am Chem Soc 105: 5925
39. Seebach D, Giovannini F, Lamatsch B (1985) Helv Chim Acta 68: 958
40. Renaud P, Seebach D (1986) Synthesis 5: 424
41. Bellamy FD, Bondoux M, Dodey P (1990) Tetrahedron Lett 31: 7323
42. Kleber HP, Seim H, Aurich H, Strack H (1977) Arch Microbiol 112: 201
43. Miura-Fraboni J, England S (1983) FEMS Microbiol Lett 18: 113
44. Miura-Fraboni J, Kleber HP, England S (1982) Arch Microbiol 133: 217
45. Seim H, Löster H, Claus R, Kleber HP, Strack E (1982) FEMS Microbiol Lett 15: 165
46. Kleber HP (1991) Metabolism of trimethylammonium compounds by *Acinetobacter*. In: Towner KJ, Bergogne-Berezin E, Fewson CA (eds) The Biology of *Acinetobacter*. Plenum, New York, p 403
47. Sih CJ (1985) WO Patent Appl 85/04900
48. Sih CJ (1988) US Patent Appl US 4751182

49. Aragozzini F, Manzoni M, Cavazzoni V, Craveri R (1986) Biotechnol Lett 8: 95
50. Kawamura M, Akutsu S, Fukuda H, Hata H, Morishita T, Kano K, Nishimori H (1987) Jpn Kokai Tokkyo Koho JP 62, 118, 899
51. Dropsy EP, Klibanov AM (1984) Biotechnol Bioeng 26: 911
52. Francalanci F, Ricci M, Cesti P, Venturello C (1987) Eur Patent Appl EP 237983
53. Bianchi D, Cabri W, Cesti P, Francalanci F, Ricci M (1988) J Org Chem 53: 104
54. Patel SS, Conlon HD, Walt DR (1986) J Org Chem 51: 2842
55. Bremer J (1962) J Biol Chem 237: 228
56. Fritz IB, Schultz SK, Srere PA (1963) J Biol Chem 238: 2509
57. Chase JFA, Pearson DJ, Tubbs PK (1965) Biochim Biophys Acta 96: 162
58. Marquis NR, Fritz IB (1965) J Biol Chem 240: 2197
59. Colucci WJ, Gandour RD (1988) Bioorg Chem 16: 307
60. Kawamoto S, Ueda M, Nozaki C, Yamamura M, Tanaka A, Fukui S (1978) FEBS Lett 96: 37
61. Ratledge C, Gilbert SC (1985) FEMS Microbiol Lett 27: 273
62. Ueda M, Tanaka A, Fukui S (1982) Eur J Biochem 124: 205
63. Claus R, Käppeli O, Fiechter A (1982) Anal Biochem 127: 376
64. Kispal G, Cseko J, Alkonyi I, Sandor A (1991) Biochim Biophys Acta 1085: 217
65. Ueda M, Tanaka A, Fukui S (1984) Eur J Biochem 138: 445
66. Miyazawa S, Ozasa H, Furuta S, Osumi T, Hashimoto T (1983) J Biochem 93: 439
67. Kozulic B, Käppeli O, Meussdoerffer F, Fiechter A (1987) Eur J Biochem 168: 245
68. Mittal B, Kurup CKR (1980) Biochim Biophys Acta 619: 90
69. Markwell MAK, Tolbert NE, Bieber LL (1976) Arch Biochem Biophys 176: 479
70. Chase JFA (1967) Biochem J 104: 1503
71. McCreary MD, Lewis DW, Wernick DL, Whitesides G (1974) J Am Chem Soc 96: 1038
72. Ouyang T, Walt DR (1991) J Org Chem 56: 3752
73. Nakayama K, Haruo H, Ogawa Y, Ozawa T, Ota T (1989) Jpn Kokai Tokkyo JP 01 222 796
74. Nakayama K, Honda H, Ogawa Y, Ozawa T, Tetsuo O (1989) Jpn Kokai Tokkyo JP 01 222 797
75. Nakayama K, Honda H, Ogawa Y, Ozawa T, Ota T (1988) Germ Offen DE 3 728 321
76. Nakayama K, Ota T (1989) Jpn Kokai Tokkyo JP 01 213 258
77. Nakayama K, Ota T (1989) Jpn Kokai Tokkyo JP 01 213 259
78. Nakayama K, Yuki O, Honda H, Okta T, Ozawa T (1989) Eur Patent Appl EP 319 344
79. Jung K, Kleber HP (1984) Wiss Z Karl-Marx-Univ Leipzig, Math-Naturwiss R 34: 293
80. Aurich H, Kleber HP, Sorger H, Tauchert H (1968) Eur J Biochem 6: 196
81. Aurich H, Kleber HP, Schöpp W (1967) Biochim Biophys Acta 139: 505
82. Kleber HP, Seim H, Aurich H, Strack E (1978) Arch Microbiol 116: 213
83. Goulas P (1988) Biochim Biophys Acta 957: 335
84. Mori N, Kasugai T, Kitamato Y, Ichikawa Y (1988) Agric Biol Chem 52: 249
85. Takahashi M, Nagasawa S, Matsuura K (1991) Ger Offen DE 4 032 287
86. Mori N, Shirota K, Kitamoto Y, Ichikawa Y (1988) Agric Biol Chem 52: 851
87. Schöpp W, Sorger H, Kleber HP, Aurich H (1969) Eur J Biochem 10: 56
88. Vandecasteele JP, Lemal J (1980) US Patent Appl US 4 221 869
89. Vandecasteele JP (1980) Appl Environm Microbiol 39: 327
90. Vandecasteele JP, Ballerini D, Lemal J, Le Penru Y (1985) US Patent Appl US 4 542 098
91. Souppe J, Haurat G, Goulas P (1987) Eur Patent Appl EP 240 423
92. Souppe J, Gisele H, Philippe G (1989) Fr Demande FR 2 621 325
93. Chocat P, Masse F, Souppe J (1989) Fr Demande FR 2 623 520
94. Donishi J, Yokozeki K (1989) Jpn Kokai Tokyo Koho JP 01 117 794
95. Schöpp W, Schäfer A (1985) Fresenius Z Anal Chem 320: 285
96. Comtat M, Galy M, Goulas P, Souppe J (1988) Anal Chim Acta 208: 295
97. Aragozzini F, Maco E, Craveri R (1986) Appl Microbiol Biotechnol 24: 175
98. Brooks DW, Kellog RP, Cooper CS (1987) J Org Chem 52: 192
99. Nakamura K, Ushio K, Oka S, Ohno A, Yasni S (1984) Tetrahedron Lett 25: 3979
100. Fuganti C, Grasselli P (1985) Tetrahedron Lett 26: 101
101. Sih CJ (1987) US Patent Appl US 4 710 468
102. Sih CJ, Chen C (1984) Angew Chem 96: 556
103. Bare G, Jacques P, Hubert JB, Rikir R, Thonart P (1991) Appl Biochem Biotechnol 28/29: 445
104. Lindstedt G (1967) Biochemistry 6: 1271

105. Abbot M, Udenfried S (1974) In: Hayaishi O (ed) Molecular mechanisms of oxygen activation. Academic, New York, p 167
106. Hayaishi O, Nozaki M, Abbott MT (1981) Oxygenases: Dioxygenases. In: Boyer PD (ed) The enzymes, Academic, New York, vol 12, p 119
107. Lindstedt G, Lindstedt S, Nordin I (1977) Biochemistry 16: 2181
108. Kondo A, Blanchard JS, Englard S (1981) Arch Biochem Biophys 212: 338
109. Lindstedt G, Lindstedt S, Midtvedt T, Tofft M (1967) Biochemistry 6: 1262
110. Lindstedt G, Lindstedt S (1970) J Biol Chem 245: 4178
111. Blanchard JS, Englard S, Kondo A (1982) Arch Biochem Biophys 219: 327
112. Punekar NS, Wehbie RS, Lardy HA (1987) J Biol Chem 262: 6720
113. Wehbie RS, Punekar NS, Lardy HA (1988) Biochemistry 27: 2222
114. Lindstedt G, Lindstedt S, Tofft M (1970) Biochemistry 9: 4336
115. Lindstedt S, Nordin I (1984) Biochem J 223: 119
116. Lindblad B, Lindstedt G, Tofft M, Lindstedt S (1969) J Am Chem Soc 91: 4606
117. Englard S, Blanchard S, Midelfort CF (1985) Biochemistry 24: 1110
118. Holme E, Lindstedt S, Nordin I (1982) Biochem Biophys Res Commun 107: 518
119. Cavazza C (1982) Ger Offen DE 31 23 975
120. Kulla H, Lehky P (1985) Eur Patent Appl EP 0158194
121. Kulla H, Lehky P, Squaratti A (1986) Eur Patent Appl EP 0195 944
122. Kulla HG (1991) Chimia 45: 81
123. Nobile S, Deshusses J (1986) J Bacteriol 168: 780
124. Nobile S, Baccino D, Takagi T, Deshusses J (1988) J Bacteriol 170: 5236
125. Nobile S, Baccino D, Deshusses J (1988) FEBS Lett 233: 335
126. Nobile S, Deshusses J (1988) Biochimie 70: 1411
127. Nobile S, Deshusses J (1988) J Chromat 449: 331
128. Watanuki M (1990) Jpn Kokai Tokkyo Koho JP 02 119 786
129. Nakayama K, Miyama M (1990) Jpn Kokai Tokkyo Koho JP 02 69 189
130. Nakayama K, Miyama M (1990) Jpn Kokai Tokkyo Koho JP 0269 188
131. Seim H, Ezold R, Kleber HP, Strack E (1980) Z Allg Mikrobiol 20: 591
132. Seim H, Löster H, Claus R, Kleber HP, Strack E (1982) FEMS Microbiol Lett 13: 201
133. Seim H, Löster H, Claus R, Kleber HP, Strack E (1982) Arch Microbiol 132: 91
134. Seim H, Löster H, Kleber HP (1982) Acta Biol Med Germ 41: 1009
135. Seim H, Kleber HP, Strack E (1979) Z Allg Mikrobiol 19: 753
136. Jung H, Jung K, Kleber HP (1989) Biochim Biophys Acta 1003: 270
137. Jung K, Jung H, Kleber HP (1987) J Basic Microbiol 27: 131
138. Seim H, Jung H, Löster H, Kleber HP (1985) Wiss Z Karl-Marx-Univ Leipzig, Math-Nat R 34: 287
139. Seim H, Löster H, Claus R, Kleber HP, Strack E (1983) DD 221 905
140. Seim H, Kleber HP (1988) Appl Microbiol Biotechnol 27: 538
141. Jung H, Kleber HP (1990) L-Carnitine synthesis by stereoselective hydration of crotonobetaine. In: Christiansen C, Munck L, Villadsen J (eds) 5th European Congress on Biotechnol 8–13 July 1990. Munksgaard Copenhagen, vol I, p 251
142. Jung H, Jung K, Kleber HP (1989) Eur Patent Appl EP 0 320 460
143. Yokozeki K, Takahashi S, Hirose Y, Kubota K (1988) Agric Biol Chem 52: 2415
144. Fukui S, Kawamura M, Akutsu S, Fukuda H (1984) Jpn Kokai Tokkyo Koho JP 61 67 499
145. Kawamura T, Iinuma S, Shinagawa S (1985) Jpn Kokai Tokkyo Koho JP 60 214 890
146. Kawamura M, Akutsu S, Fukuda H, Hata H, Morishita T, Kano K, Nishimori H (1986) Jpn Kokai Tokkyo Koho JP 61 234 788
147. Kawamura M, Akutsu S, Fukuda H, Hata H, Morishita T, Kano K, Nishimori H (1986) Jpn Kokai Tokkyo Koho JP 61 234 794
148. Kawamura M, Akutsu S, Fukuda H, Hata H, Morishita T, Kano K, Nishimori H (1986) Jpn Kokai Tokkyo Koho JP 61.271 995
149. Kawamura M, Akutsu S, Fukuda H, Hata H, Morishita T, Kano K, Nishimori H (1986) Jpn Kokai Tokkyo Koho JP 61 271 996
150. Whitesides GM, Wong CH (1985) Angew Chem 97: 617

Convective Drying of Bacteria
I. The Drying Processes

L. C. Lievense[1] and K. van 't Riet[2]
Food and Bioprocess Engineering Group, Wageningen Agricultural University,
P.O. Box 8129, 6700 EV Wageningen, The Netherlands

Because of the large number of applications of starter bacteria in the food industry, it is important to obtain stable, highly viable bacterial cultures. This review describes the possibilities of producing these cultures in a dried form. Freeze-drying is economically unattractive for producing large quantities of dried bacteria. Therefore, three convective drying methods are emphasized: spray drying, fluidized-bed drying and spray granulation. Important parameters which can influence the inactivation of the bacteria during these drying processes are considered.

[1] Present address: Unilever Research Laboratory, P.O. Box 114, 3130 AC Vlaardingen, The Netherlands
[2] To whom correspondence should be addressed

Advances in Biochemical Engineering/
Biotechnology, Vol. 50
Managing Editor: A. Fiechter
© Springer-Verlag Berlin Heidelberg 1993

1 Introduction

Since ancient times, bacterial starter cultures have played an important role in the production of food and feed. Modern processes require cultures with well-defined properties in order to generate end products with a consistent quality. Therefore, it is possible that techniques applied originally, such as subculturing or growth of bacteria naturally present, can no longer be used. It is then necessary to find suitable preservation methods for obtaining 'ready to use' bacteria. In a preservation method such as freezing and/or drying, the bacteria are brought to an 'anabiosis state'. In this state the bacterial metabolism is reversibly reduced to an extremely low level [1] and the bacteria can be preserved for a longer time.

Normally, bacteria are used in the food industry as stock cultures for the preparation of bulk starters. A disadvantage of using stock cultures is the necessity of culturing the large volume that is required to inoculate the process liquid from this stock. Such a production is often accompanied by the risk of contamination. It is preferable, that the formulations used contain enough bacteria to inoculate the process liquid directly ('direct to vat' cultures). This reduces the risk of contamination and constant culture characteristics and quality can be guaranteed. Frozen and freeze-dried formulations are used for this purpose. The main advantage of using dried instead of frozen bacteria is the lower costs in transport and storage of the cultures.

The main disadvantages of the dried cultures are the considerable inactivation of the culture during drying and the poor shelf life of the product under uncontrolled conditions. Often, the producer must add many more bacteria and so give a far higher activity than guaranteed. Under certain conditions, the inactivation during drying and storage can be too great to develop a commercial process. However, the large commercial interest in bacterial starters explains the continuing interest in the drying of these cultures. At the moment the disadvantages of dried cultures are hindering their application but these disadvantages can be overruled when the inactivation during drying and storage is clearly understood and reduced.

Since the beginning of this century a number of researchers have investigated the drying of microorganisms. Rogers [2] was probably the first author to write about this subject. Most publications deal with the freeze-drying of bacteria in order to obtain stable stock cultures. Attention was also given to the convective drying of yeast *(Saccharomyces cerevisiae)*. Reviews on freeze-drying were written by Heckly [3–6], Bousfield and MacKenzie [7] and Ashwood-Smith [8]. Reviews on convective yeast drying were written by Josic [9] and Beker and Rapoport [1]. This review places emphasis on the convective drying of bacteria. It will not cover the freeze-drying of bacteria nor the drying of yeasts. Freeze-drying and convective yeast drying is only referred to, where it gives more insight in the convective drying of bacterial cultures.

2 Preservation Methods

To produce a stable bacterial culture, a number of methods are available. First of all, one has to distinguish between typical laboratory methods and methods that can be applied to large scale production. Typical laboratory methods include: 1) immersing a bacterial culture in oil (e.g. paraffin oil), 2) air drying of the cultures in gelatin or agar, 3) adsorption-desiccation on filter-paper or on predried plugs of starch, peptone or dextran, or on sand, soil, kieselguhr, pumice stone, porcelain, or silica gel [10–17]. With these methods, many bacteria can be preserved successfully, for months or even years. An essential difference between the requirements for laboratory and industrial use is that for the former the lowest survival percentage is satisfactory, as one is only interested in the possibility of obtaining a fresh culture after inoculation with the stored bacteria.

For industrial use, large quantities of active bacteria are required, particularly when direct inoculation of the process liquid is applied. The above mentioned laboratory methods are not suitable for this purpose because of their complexity, the additives needed, and the low survival rate for most of the applications. Methods which can be used for industrial preservation are: 1) subculturing, 2) storage of the culture in a frozen form, 3) storage of the culture in a dried form. These preservation methods will be discussed below.

Subculturing is a traditional method of preserving bacteria through periodic transfer to fresh media [12]. Although this method is labour intensive, it is still commonly used in the traditional food industry. In a bakery, for example, the starter culture is often obtained from the preceding batch [18, 19]. However, this method entails the risk of genetic instability and contamination [16]. When subculturing is used in industry, often a frozen stock culture is kept in reserve.

Freezing is the most frequently used preservation method for starter cultures [12, 16, 20, 21]. Generally, skim milk, glycerol, lactose, and/or a calcium salt are used as protective additives. At temperatures above $-30\,°C$, there is still a risk of genetic instability and loss of viability. Therefore, very low temperatures (down to $-196\,°C$), which can increase the stability of a culture considerably, are used [12, 21–25]. To save freezing costs and to obtain a product with a high concentration of viable cells, cultures are usually concentrated by centrifugation [20] before freezing. The costs of transport and storage of deep-frozen cultures are a major drawback.

Freeze-drying is also a widespread preservation technique. From the freeze-drying reviews (see Introduction) is it clear that survival and stability of the cultures after freeze-drying can differ widely. A number of factors influence this survival and stability. Most of these factors are also important in the convective drying of bacteria and are discussed in Part II of this review [26]. Most investigators had the sole objective of initiating growth readily on a subculture without optimization of the survival. Yet it appears that, providing suitable protective media and optimal process and storage conditions are applied, the survival rate can be high (40–80%) for a number of dehydration-resistant strains [27–32]. Therefore, freeze-drying of starter cultures can be used as a commercial process and consequently a number of freeze-dried cultures are on the market [22–24]. Nevertheless, due to the high costs and complexity of the process itself, freeze-drying is generally not considered an attractive method for the preparation of large quantities.

When freeze-drying is compared with convective drying, the former has some major disadvantages. Not only is the sublimation of water more energy consuming than evaporation, but also the investment costs for a freeze-drying plant are higher [33, 34]. Due to the vacuum required, it is quite complicated to apply freeze-drying in a continuous process. It is relatively easy, however, to operate a batch freeze-drying process aseptically. From a biological viewpoint it is important to note that not only the dehydration process but also the freezing process can be responsible for a considerable inactivation [3, 6]. However, it is the high cost of the classical freeze-drying process which is the main reason for the continuing search for alternative drying methods.

3 Preservation by Convective Drying

Several drying methods can be used for the preservation of microorganisms. Among the convective drying techniques for bacteria, only three are of real interest: spray drying, fluidized-bed drying and a combination of both, spray granulation. In Fig. 1 the principle layout of each drying method is given. Detailed technical descriptions are to be found in various literature sources [35–41]. In the next sections, the application of the three methods and the influence of a number of drying process parameters on the inactivation of the bacterial cells, will be discussed.

3.1 General Considerations

It is generally assumed that thermal inactivation is the main reason for inactivation during convective drying. Recently it was showed that the inactivation of L. plantarum during fluidized-bed drying was caused by thermal inactivation and by the dehydration itself [42, 43]. These two inactivation mechanisms can occur simultaneously during a convective drying process. Other authors [3, 7–9, 44–46] have also described a form of dehydration inactivation. Usually, the term 'critical water concentration' was used, below which the bacterial cell will be inactivated. It is likely, however, that dehydration inactivation will occur not at a single water concentration but within a water concentration range [43]. This may be due to biological variations in the cell population. Throughout this review it is assumed that the inactivation during convective drying can be caused by two mechanisms, thermal inactivation and dehydration inactivation.

At high drying temperatures, thermal inactivation will play a substantial role. It is important to realize that, in general, the heat resistance of bacterial cells increases with decreasing water concentrations [47, 48]. Daemen [49] found that the heat resistance at 50 °C of Serratia marcescens mixed with milk powder was increased 17 times when the water concentration was decreased from 10 to 0.07 kg of water per kg of the solids. Zimmermann [50] reported a two-fold increase in the heat resistance at 50 °C of S. cerevisiae when the water concentration was decreased from 2.3 to 0.3 kg kg^{-1}. The heat resistance of L. plantarum which had been immobilized in potato starch at 50 °C increased nine-fold, when the water

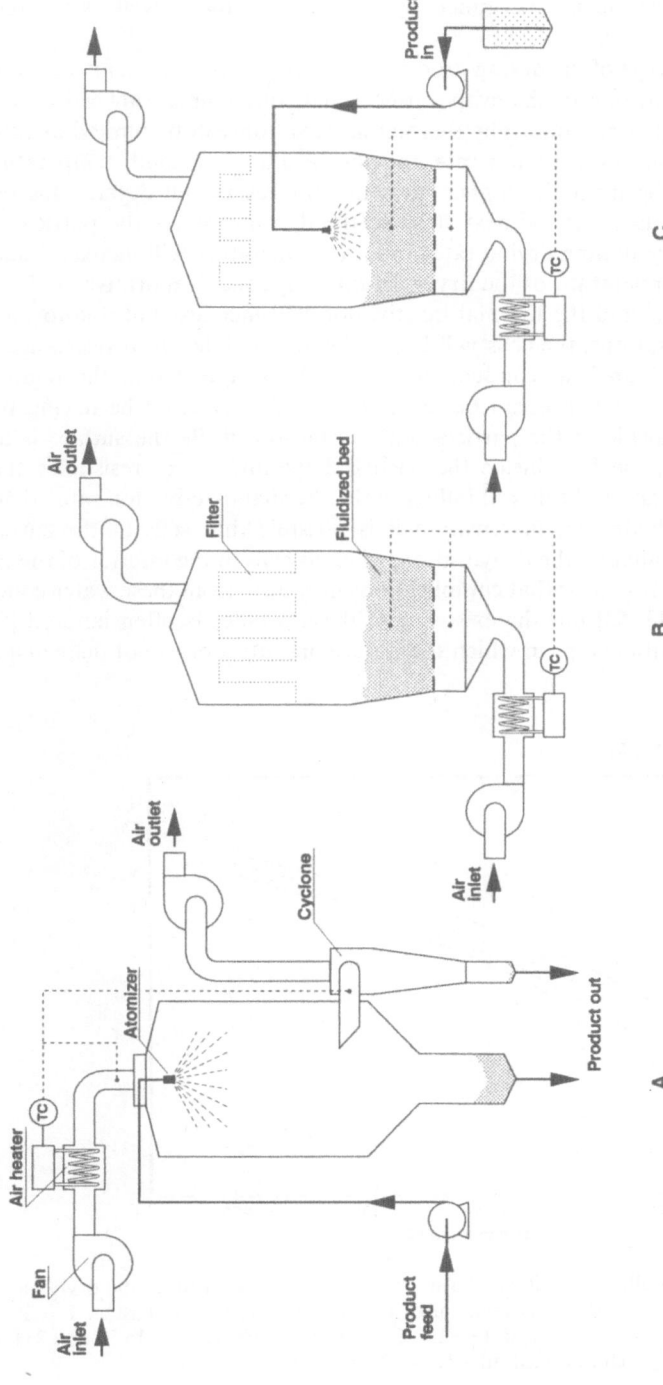

Fig. 1A–C. The principle layout of the three drying methods. **A** Continuous spray drying, **B** batch fluidized bed drying, and **C** batch spray granulation. The temperature of the inlet air can be controlled by measurement of the inlet air, the outlet air, or the bed temperature

concentration was decreased from 0.8 to 0.15 kg kg^{-1} [42]. Hence, the overall thermal inactivation is determined by the temperature as well as the moisture history of the system.

In the first stage of the drying process the particle surface remains wet (constant drying rate) and due to the evaporating water, the temperature will not exceed the 'wet-bulb' temperature. The thermal inactivation will be limited at this stage because the high evaporation rate and the resulting wet-bulb temperature will protect the cells from the higher air temperatures in the dryer. This effect is quantified in Fig. 2. At the next stage in the drying process the particle surface will become dry (falling drying rate) and the temperature will increase maximally to the inlet temperature of the dryer. In this stage the evaporative cooling is no longer available and the thermal inactivation will increase, but due to the lower water concentrations, the cells will have (generally) higher heat resistance.

Due to the high heat conductivity, it can be assumed that the temperature profile inside the particle can be neglected [51, 52]. During the drying process, however, the inside of the particle can still be wet, while the surface is already dry. Therefore, the cells inside the particle have lower heat resistance than the cells at the surface and this will influence the thermal inactivation rate. Moreover, due to the dehydration inactivation it is possible that cells at the surface are already inactivated by the dehydration, whereas cells in the interior of the particle are unaffected. It is clear that complications may arise from these water concentration profiles [43, 53] but the existence of these profiles is often ignored [54, 55]. As a result, relationships in which the water concentration is not defined as being

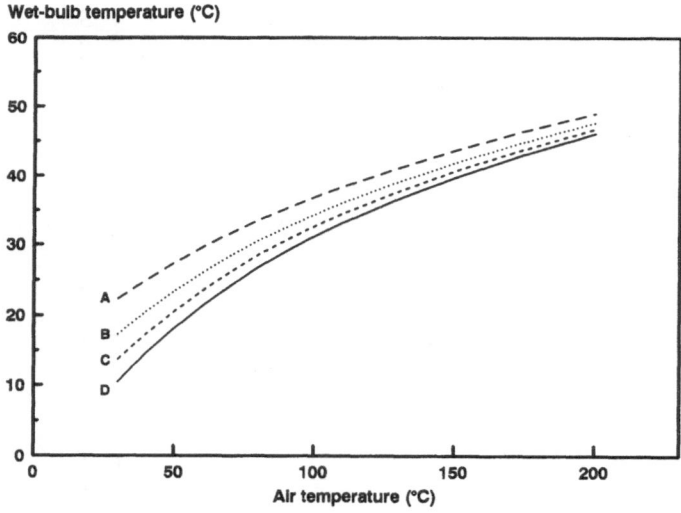

Fig. 2. The wet-bulb temperature as a function of the temperature of the drying air [82] at different conditions of the external air used for drying. External air with *A*. T = 25 °C and RH (relative humidity) = 70%, *B*. T = 20 °C and RH = 50%, *C*. T = 15 °C and RH = 30% and *D*. completely dehumidified air (RH = 0%)

local or overall, must be considered critically. This is especially important in the early stage of the drying process where water concentration profiles may be steep.

When survival data of bacterial cells are compared, it is important to realize that this is only meaningful, when the same bacteria are dried and rehydrated under the same conditions. Furthermore, the survival data have to be compared together with the final water concentration. (This water concentration is preferably described in unambiguous terms. Throughout this work the unit as water per solids 'kg kg^{-1}' will be used. The expression '%-moisture' is considered as 'kg H$_2$O per kg total $\times 100\%$'.) In most literature sources these requirements are not fulfilled and, therefore, a quantitative comparison can seldom be made.

In the next sections the literature data are outlined in view of the simultaneously occurring thermal and dehydration inactivation. For each drying method the discussion will be started with an overview of the process parameters which influence the moisture-temperature history of the cells. Next, several process designs will be discussed. This is followed by a review of several other factors which can be important for obtaining high quality starter cultures in a dried form and possible advantages/disadvantages of each drying method. The main parameters of the drying processes that can influence survival during drying are summarized in Table 1.

Table 1. Several parameters of the convective drying processes which can influence the survival of bacteria upon drying

	Spray drying	Fluidized-bed drying	Spray granulation
	continuous	continuous/batch	continuous/batch
Thermal inactivation			
General	residence time (distribution) drying rate/time inlet temperature mass ratio air/feed	residence time (distribution) drying rate/time inlet temperature air velocity bed porosity T-profile bed	residence time (distribution) drying rate/time inlet temperature air velocity bed porosity T-profile bed
Particle/Granule size (distribution)	agglomeration nozzle type/shear (P atomization) density/concentration feed surface tension/viscosity feed	agglomeration (stickiness granules) granulation process	agglomeration (stickiness granules) nozzle type/shear (P atomization) spraying rate/time
Dehydration inactivation			
	dehydration rate final water concentration	dehydration rate final water concentration	dehydration rate final water concentration

3.2 Spray Drying

Much research has been performed on spray drying of bacteria. In view of the thermal inactivation, the main advantage mentioned is the rapidity of drying [37, 56, 57], due to the high specific drying surface of the spray. This results in a short residence time for the bacterial cells in the spray dryer (20–40 s; [36, 58]) and this will limit the thermal inactivation. In a spray drying process, high inlet temperatures are necessary precisely because the specific heat of evaporation must be supplied in such a short time. As a result of this, significant thermal inactivation of the cells can occur, despite the short residence time.

Because of the evaporative cooling in the first part of the drying process, the survival of bacterial cells during spray drying is strongly correlated to the outlet temperature and not directly to the inlet temperature of the dryer. This relation is confirmed by most of the relevant literature. The highest survival rate is found at the lowest outlet temperatures. Obviously, low outlet temperatures will result in less thermal inactivation but possibly also in higher residual water concentrations which can influence the survival as well. Substantial thermal inactivation of the cells can be avoided by choosing the correct mass ratio of drying air to liquid feed, so that at the outlet of the dryer the air will be cooled by the evaporation of water.

The moisture-temperature history of a system is influenced significantly by the particle size. The particle size depends on three physical properties of the cell suspension: viscosity, density and surface tension. The viscosity is the only parameter that can be changed considerably by a change of the cell concentration [9]. A higher solid concentration of the cell suspension will thus result in larger particles [56, 59, 60]. With increasing particle size the drying time will also increase. This can decrease the survival rate because the contact time of the particles with hot air (required to reach a certain final water concentration) will increase. Therefore, advising the use of solid concentrations as high as possible in the feed to the spray dryer as a means of saving energy costs [36, 37] is not to be recommended if a high survival rate is required.

The particle size can also be influenced by the nozzle type and the atomization pressure [37]. A decrease in the survival rate of *Lactobacillus bulgaricus* and *S. thermophilus* (from 19.5 to 17% and 17 to 11%, respectively) due to larger particle size at lower atomization pressure (250 vs. 110 kPa) was reported by Metwally et al. [57]. Kim and Bhowmik [61] found, in spray drying experiments on yogurt, that the survival of *L. bulgaricus* and *S. thermophilus* slightly increased (from 0.2 to 0.8% and 1.5 to 2.6%, respectively) with decreasing atomization pressure (from 200 to 100 kPa). The final water concentrations reached in these experiments were not reported. At a fixed air flow rate, a smaller particle size at high atomization pressure gave a longer residence time [62] and thus a higher thermal and dehydration inactivation. Comings et al. [56] used a specially designed jet spray dryer for the drying of *S. marcescens*. In this device [41, 63] the primary drying air moves at near-sonic velocity (290 m s^{-1} at 0.17 m from nozzle). Very fine atomization results with extremely short drying times in the range of milliseconds [56]. When *S. marcescens* was dried with this device an average of survival of

48% at 0.05 kg kg^{-1} solids was found, significantly higher than found in other work with this bacterium (2% survival at 0.06 kg kg^{-1}; [49, 64]).

Another item which can influence the survival is the design of the spray drying process. Several designs are described in Masters [36] and Filková and Mujumdar [37]. Also a combination of spray drying as the first stage and fluidized-bed drying as the second stage is mentioned, a design often used for the production of milk powder. Such a design can be used for the drying of bacterial cells and yeasts [1, 19, 41, 65]. In this design the product is dried to a relatively high water concentration $(0.1-0.2 \text{ kg kg}^{-1})$ in a spray dryer. Then the powder is cooled immediately and further dried in a fluidized bed, until the desired water concentration is reached. This design combines the intensive initial evaporation of spray drying with a better temperature controlled fluidized bed. Quantitative survival data for bacterial cultures dried with this layout are not reported. In addition, a combination of spray drying and vacuum drying for the drying of a mixed starter culture was reported with an overall survival of 36% at 0.05 kg kg^{-1} [20].

Several other factors of the spray drying process can be important for obtaining a high quality dried starter culture. One can expect that the shear in the nozzle influences the survival rate but there was no effect of 'nozzle-shear' on the survival of *Streptococcus lactis* [66] or *S. cerevisiae* [58, 67]. Another factor that can influence the survival rate are wall deposits in the spray dryer. These deposits increase the residence time and thus the (thermal) inactivation of the bacterial cells. Wall deposition will be influenced by the water concentration of the powder [56] and thus the drying rate.

Several advantages/disadvantages of the spray drying process can be mentioned. An advantage of the spray drying process is that the bacterial cells can be dried as a suspension, without the need of support material. The maximum concentration of the bacterial suspension is limited by the fact that the feed to the dryer must be pumpable and for the reasons of particle size mentioned above. This means that an excess of water has to be evaporated and so the spray drying process is relatively energy consuming.

A disadvantage can be the usually dusty powder that is obtained after drying [37, 68]. A secondary drying step, consisting of a vibrating fluidized bed can be used for agglomeration of the spray dried powder [36]. Particle agglomeration occurs during the fluidized-bed drying stage and prevents dusty powders. The usually good reconstitution properties of spray dried products can be regarded as an advantage while the low bulk densities of these powders increase the packaging and storage costs.

As mentioned above, a proper choice of the flow rate and temperature of the inlet air and the flow rate of water that has to be evaporated (and thus the flow rate of feed pumped to the nozzle), together with particle size and size distribution, will minimize the thermal inactivation. The right settings for these variables are mostly found by trial and error, because it is difficult to calculate them in advance. The requirement of performing such experiments, the relatively difficult control during the drying process itself, and the complicated modelling/ optimization of the process can be regarded as disadvantages of the spray drying process.

3.3 Fluidized-Bed Drying

When compared with spray drying, publications on drying bacterial starter cultures with a fluidized bed are rare. Most of the fluidized-bed work is related to yeast drying. In a fluidized-bed dryer the drying time can be much longer than in a spray dryer (e.g. 60 min vs. 30 s). Because the dimensions of the drying apparatus and the required ratio air flow/moisture flow to reach a certain final water concentration are no longer coupled, the residence time in the fluidized bed can be chosen freely. This also means that the air inlet temperature can be controlled without influencing the minimal obtainable water concentration after drying.

The free choice of residence time makes it possible to use relative low air temperatures, which will help to minimize thermal inactivation. The thermal inactivation can also be minimized by controlling the product temperature, instead of the inlet temperature of the bed (Fig. 1). For example, during the drying process the inlet air temperature can be varied from 160 °C at the start of the drying process to 30 °C at the end, while the product temperature remains at 30–35 °C [1, 69, 70]. The process design can help to minimize the thermal inactivation. For example, yeast can be dried with a constant air inlet temperature of 80 °C at the first stage where charges of 1500 kg compressed yeast (2.3 kg kg^{-1}) are dried in 45 min to water concentrations of 0.4 kg kg^{-1}. At the second stage, where water concentrations of 0.1 kg kg^{-1} are reached in 45 min, the air temperature can be controlled in such a way, that the product temperature never exceeds 38 °C [71]. In the final stage of the process the air humidity can also be controlled, thereby limiting the dehydration and thus the inactivation of the yeast cells [1, 72]. These control schemes can be used to minimize both thermal and dehydration inactivation. By accurate control of temperature and water concentration, in relation to particle diameter, an optimum between inactivation and drying time can be reached.

In addition, combinations of processes, such as the spray and fluidized-bed drying mentioned above or (flash pneumatic) conveyer drying followed by fluidized-bed drying, are used [1, 72]. Due to the usually large distribution in residence time, a continuous fluidized-bed drying process can be disadvantageous for the drying of microorganisms. At higher drying temperatures, a distribution in residence time can cause significant thermal inactivation. In a batch-operated process there is no variation in residence time but such a process is less attractive from an economical view point. Generally, series of batch-operated fluidized beds are used.

A procedure that can improve the survival rate after the fluidized-bed drying process is osmotic predehydration. Before drying, the water concentration of a yeast suspension is generally lowered by filtration. An additional osmotic dehydration step with a salt solution [68] can be carried out before or during this filtration. Pomper [73] found that osmotic dehydration with divalent metal salts solutions (Mg^{2+}, Ca^{2+}) gave a 20–30% higher survival rate than sodium or potassium salt solutions. Additionally, highly concentrated solutions of sugar (saccharose, glucose fructose), polyalcohol (sorbitol, innositol) or glycerol can be used for this purpose [55]. In addition, with these solutions a high survival rate after fluidized-bed drying was found. Bacterial cells cannot be readily filtered, and most often centrifugation or membrane filtration is used as the concentration step. If, at this stage, the cells are not immobilized, then osmotic dehydration must be carried out in the concentrated cell suspensions. Relatively large amounts of osmotic dehydration media are needed in such a procedure. Further problems can arise in later separation steps due to the high viscosity of the osmotic dehydration solutions. Moreover, it is not clear if a positive effect on the survival is caused by the osmotic dehydration step itself, the shortening of the drying time due to the lower initial water concentration, or by specific protective interaction of the compounds with the cells [26].

Several additional advantages/disadvantages of the fluidized-bed drying process can be mentioned. A general problem during fluidized-bed drying can be the stickiness of the granulated material. This can influence the survival rate after fluidized-bed drying. Sticky particles agglomerate easily, which can result in a substantial increase in particle size, inhomogeneous beds, and decreasing drying rates. This problem can possibly be avoided by adjusting the initial water concentration, using the disintegration forces of a vibrating grid under the bed [1], a right choice of the support material, and/or the use of fluidizing agents [50]. Another drawback is that only granulatable materials can be dried. Bacterial cells are not readily obtained in a granulatable form and therefore, support materials have to be used. One possibility is to mix the cells with a support material such as starch [42, 43, 53, 72] or wheat bran [75] and to extrude the paste formed. Gel-like materials are also used, such as xanthan gum, carrageenan or alginate [55, 76, 77]. Unfortunately, these materials can be considered unacceptable in the (food) substance to be inoculated. With regard to a spray granulation process (see below), Hill [33] suggests using a support material which fits a particular purpose. Milk powder or lactose is mentioned for milk starters, maltodextrin for sausage starters, and rye or wheat flour for bread starters. Concentration, mixing and granulation are unavoidable processes preceding the fluidized-bed drying process. In these preceding steps, it is possible that inactivation of the cells may occur. However, there is no known literature about this effect.

Another disadvantage can be the poor reconstitution properties of the fluidized-bed-dried product. Usually, the product is less porous when compared to a spray or freeze-dried product and rehydration times can be relatively long [1, 68]. To facilitate the rehydration (and dehydration) process of dried yeast, the paste-like material is extruded as very small particles with diameters of 0.2 mm [68]. After drying, particles of 0.1 mm diameter remain which are small enough for direct addition into the dough, without the need of a separate rehydration step.

An advantage of the fluidized-bed process, compared to the spray drying process, is that it can be modelled relatively easily. The insights into the influence of the parameters on the drying and inactivation process, obtained with the modelling and simulation, can be readily implemented. The measurements and calculations practiced with a laboratory or pilot plant drying installation can be translated to an industrial scale process.

3.4 Spray Granulation

Spray granulation is a relatively new technique which is frequently used in the pharmaceutical industry. The terms 'fluid-bed granulation' or 'fluid-bed spray drying' are also used. As far as we are aware, only three research groups have reported the use of this technique as a drying method for bacterial cells [33, 50, 70, 77–79].

Spray granulation can be defined as a particle-forming process by which a liquid feed containing solids is converted to a granular state. This is achieved by spraying the feed into a fluidized-bed of previously formed granules [38, 39]. In the first stage of the process the cell suspension is sprayed on a fluidized bed consisting of, for example, a starch powder. The fine starch particles in the bed then agglomerate together with the cells to form larger granules. In the second stage of the process these granules can be dried further to the desired final water concentration.

The thermal-inactivation rate is determined by the moisture-temperature history of the cells which in turn, will be influenced by the size of the formed granules. The granule size can be controlled by a number of variables of which the droplet size of the spray is the most important [38]. Apart from granule size, two other factors should be mentioned. Firstly, the bacterial cells sprayed on the support material will dry at a fast initial rate because of water absorption by the support material. As a result, the heat resistance of the cells will increase rapidly. Secondly, the spray granulation process has the same control potential as the fluidized-bed drying process, and by optimum control, the thermal inactivation can be minimized.

The differences in survival rate found by several authors who used the spray granulation process (see Table 2), are certainly influenced by the choice of species, but also differences in process control can be responsible. Hill [33, 70, 78] adjusted the inlet temperature during drying (from 62 to 38 °C), Zimmermann [50] used a fixed inlet temperature (78 °C) whereas Roelans and Taeymans [77] did not mention temperature control. Zimmermann [50] measured a temperature profile in the bed from 75 °C at the bottom to 30 °C at the top of the bed. Obviously, with these temperatures, thermal inactivation of the bacteria can occur. No quantitative relation between inlet temperature and survival can be found from the reported data [50] but it can be shown that the highest survival rate is obtained at the lowest air inlet temperatures. This is confirmed in Zimmermann and Bauer [79] where only inlet temperatures below 50 °C were applied. The differences in survival could also have been caused by other variations in process operation, such as spraying time, second-stage drying time, air humidity control and final water concentration.

In the spray granulation process the formation of granulates prior to drying is omitted, while the control possibilities of fluidized-bed drying still exists. Also, in this process, the presence of a support material in the end product can be a disadvantage. Nevertheless, it is clear that the combined advantages of spray drying and fluidized-bed drying make the spray granulation process an attractive alternative. As with the spray-drying process, a model description for spray granulation is not easily obtainable [39], however, an empirically based control is possible. For this control the remarks made for fluidized-bed drying are again relevant.

3.5 Choosing a Drying Process

When one is interested in the production of a dried bacterial starter concentrate at low cost, convective drying is unavoidable. One of the three methods mentioned may be chosen. In Table 2, a summary is given of the references we have found in which quantitative survival data are reported for the convective drying of bacteria. In Fig. 3 the survival data of the lactic acid bacteria in Table 2, are plotted as a function of the final water concentration. In this figure no clear correlation can be seen between survival and water concentration. Large variations in drying conditions and other factors [26] diminish this correlation. Nevertheless, Fig. 3 does reveal some trends.

Hill [33, 70, 78] obtained the most promising results with a spray granulation process. Zimmermann [50] and Zimmermann and Bauer [79] used the same process and found a significantly lower survival. However, the reported water concentrations are also very low. Apart from the results of Metwally et al. [57] who used intermediate outlet temperatures (around 70 °C), the results of spray drying with a high outlet air temperature (above 75 °C) are located significantly lower than the results with low outlet temperatures (below 50 °C). Also, Metwally et al. [57] found a significantly lower survival rate at higher outlet temperatures. The fluidized-bed survival data are located in the intermediate range. In the literature, survival data vary widely both within and between the three drying methods and a rational choice based on Table 2 or Fig. 3 cannot be made. Unfortunately, it is not possible to give a straightforward recommendation for one of the three aforementioned processes.

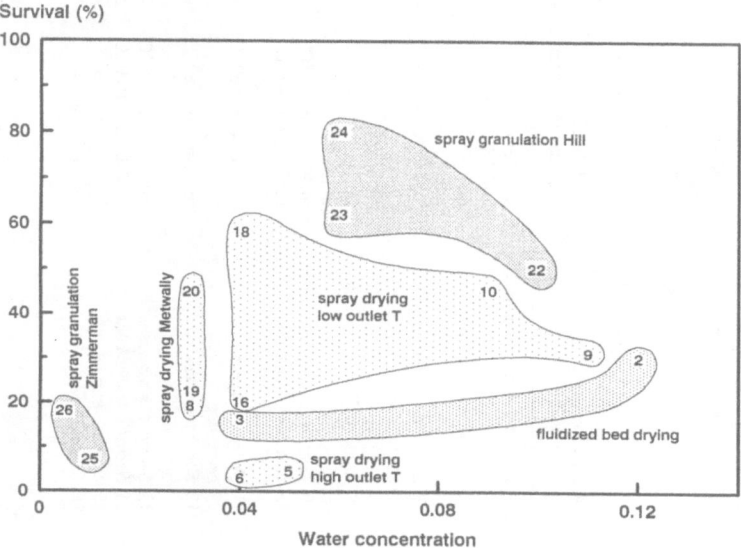

Fig. 3. An overview of the results found with lactic acid bacteria dried by different methods (the numbers refer to Table 2). Water concentration in the number of kg of water per kg of solids

Table 2. Convective drying of bacteria (only references which mention quantitative survival data)

Drying method	Micro-organism	Drying conditions	Medium-Protectant/Support	Residual H_2O concentration (kg kg^{-1})[a]	Survival (%)[ab]	Reference	Comments
Fluidized-bed	1 *Leuconostoc oenos*	in: 30–40 °C 25–75 min	water/soluble starch	0.1–0.4	1–10	74	
	2 *Lactobacillus plantarum*	in: 30 °C 60 min	potassium phosphate buffer/native potato starch	0.12	30	42, 43	survival: acid production
	3 *Lactobacillus plantarum*	in: 30–40 °C 240 min	culture medium/wheat bran	0.04	17–19	75	
	4 *Saccharomyces cerevisiae*	in: 120–35 °C product: 35 °C 8 min	sorbitan mono-stearate/no support	0.07	75–91	69	survival: gas production
Spray-drying	5 *Lactobacillus acidophilus*	in: 170 °C out: 75–80 °C	NFMS	0.04–0.06	0.3–8	60	
	6 *Lactobacillus acidophilus*	in: 170–180 °C out: 75–90 °C	skim milk	0.03–0.05	0.4–3.5	80	
	7 *Lactobacillus acidophilus*	in: 140 °C out: 70 °C	skim milk		13–18	81	±6% higher survival with Na-glutamate or Na-citrate
	8 *Lactobacillus bulgaricus*	in: 190–200 °C out: 71 °C	–	0.03	19.5	57	
	9 *Lactobacillus plantarum*	in: 90 °C out: 42 °C	native potato starch	0.11	31	43	survival: acid production
	10 *Mixed starter culture*[c]	in: 70 °C out: 42 °C	neutralized culture medium	0.09	45	20	survival: total number of bacteria
	11 *Saccharomyces cerevisiae*	in: 190 °C out: 60 °C	glycerol	–	0.02	76	
	12 *Saccharomyces cerevisiae*	in: 180 °C out: 60 °C	water	0.09	1.6	67	
	13 *Serratia marcescens*	in: 110 °C out: 65 °C	–	0.04	43	59	
	14 *Serratia marcescens*	in: 260 °C out: 60 °C	dextrin and ascorbic acid	0.05	86[d]	56	
	15 *Serratia marcescens*	in: <100 °C out: 50 °C	skim milk	0.06	2	49, 64	
	16 *Streptococcus cremoris*	in: 80 °C out: –	NFMS	0.04	18	21	jet spray dryer

	Drying conditions	Medium	a_w	Survival	Survival	Remarks
17 *Streptococcus lactis*	in: 60–180 °C out: –	peptone and NaCl	–	<1	66	
18 *Streptococcus lactis*	in: 80 °C out: –	NFMS	0.04	53–63	21	
19 *Streptococcus lactis*	in: 190–200 °C out: 71 °C	–	0.03	22	57	
20 *Streptococcus thermophilus*	in: 190–200 °C out: 71 °C	–	0.03	45	57	
Spray granulation						
21 *Lactobacillus acidophilus*	in: – bed: 45–50 °C 45 min total	MRS-medium/milk powder	–	24	77	a_w of end product: 0.25
22 *Lactobacillus brevis*	in: – bed: 30 °C 15 min spraying + 30 min drying	culture medium/ wheat flour	0.1	50	78	
23 *Lactobacillus brevis*	–	–/wheat flour	0.06	62	33	
24 *Lactobacillus brevis*	in: 60–40 °C bed: 20–35 °C 10 min spraying + 20 min drying	–/rye flour	0.06	80	70	
25 *Lactobacillus casei*	in: 78 °C bed: – [e] 40 min spraying in 65 min total	MRS-medium/lactose	0.01	7	50	±10% higher survival with protectants
26 *Lactobacillus casei*	in: 30–48 °C bed: 25–39 °C 150 min spraying in 160 min total	MRS-medium/lactose	0.005	18	79	±20% higher survival with NaCl as support

[a] Approximate values, especially those obtained by estimation from data in figures; [b] Survival measured with CFU-count, unless mentioned otherwise; [c] Mixed starter culture including *Streptococcus diacetilactis*; [d] Highest survival reported; average survival 48%; [e] See text spray granulation

in = inlet air temperature
out = outlet air temperature
– = not mentioned
NFMS = reconstituted Non Fat Milk Solids

Table 3. Several advantages and disadvantages of the convective drying processes

	Spray drying	Fluidized bed drying	Spray granulation
Support needed	no	yes	yes
Additional process steps	no	yes	no
Energy consumption	−	+	−
Modelling	−	+	−
Control	±	+	+
Reconstitution product	+	−	±
Dustiness product	−	+	+
Density product	−	+	±

+ = advantageous
− = disadvantageous

Several advantages/disadvantages of the three drying processes are summarized in Table 3. The most important advantage of spray drying is that no support material is needed. Alternatively, fluidized-bed and spray granulation drying offer the freedom of choosing the drying time and good control possibilities. From the perspective of survival, when the difference in control possibility is eliminated by advanced spray drying techniques, the choice between the three drying processes becomes trivial. With optimal control, high thermal inactivation can be avoided. In some cases, and depending on the application, the other advantages/disadvantages may be important. It is likely that combined processes can help to optimize survival.

If the dehydration rate is found to be important [26], then the drying process has to be chosen accordingly. However, dehydration inactivation of the cells is inevitable during drying and furthermore, the dehydration inactivation will not be influenced by the choice of the drying process but is determined by other factors which are discussed in Part II of this review [26].

Acknowledgement: We would like to thank Dr. H. H. Beeftink, Mr. G. Meerdink, Dr. M. R. Smith, Dr. E. A. Tudor and Mr. M. H. Zwietering for their helpful critical comments in the preparation of this manuscript. The financial support of Gist-brocades NV, Delft, The Netherlands, is gratefully acknowledged.

4 References

1. Beker MJ, Rapoport Al (1987) In: Fiechter A (ed) Advances in biochemical engineering/ biotechnology, vol 32 Springer, Berlin Heidelberg New York, p 127
2. Rogers LA (1914) J Infect Diseases 14: 100
3. Heckly RJ (1961) In: WW Umbreit (ed) Advances in applied microbiology. Academic, NY, p 1
4. Heckly RJ (1978) In: Crowe JH and Glegg JJ (eds) Dry biological systems. Academic, NY, p 257
5. Heckly RJ (1978) Cryobiology 15: 654
6. Heckly RJ (1985) Develop industrial microbiol 26: 379

7. Bousfield IJ, MacKenzie AR (1976) In: Skinner FA and Hugo WB (eds) Inhibition and inactivation of vegetative microbes. Academic, London, p 329
8. Ashwood-Smith MJ (1980) In: Ashwood-Smith MJ and Farrant J (eds) Low temperature in medical biology. Pitman Medical, Turnbridge Wells, p 219
9. Josic D (1982) Lebensm-Wiss-Technol 15: 5
10. Lal M, Tiwari MP, Sinha RN, Ranganathan B (1976) J Food Sci Techn India 13: 266
11. Sleesman JP, Leben C (1978) Plant disease reporter 62: 910
12. Gherna RL (1981) In: Gerhardt P (ed) Manual of methods for general bacteriology. Am Soc Microbiol, Washington, p 208
13. Bousfield IJ (1984) In: Kirsop BE and Snell JJS (ed) Maintenance of micro-organisms a manual of laboratory practice. Academic Press, London, p 63
14. Leben C, Sleesman JP (1984) Plant disease 66: 327
15. Snell JJS (1984) In: Kirsop BE and Snell JJS (ed) Maintenance of micro-organisms, a manual of laboratory practice. Academic, London, p 11
16. Barbour EA, Priest FG (1986) Letters in appl microbiol 2: 69
17. McEldowney S, Flechter M (1988) Letters in appl microbiol 7: 83
18. Spicher G (1983) In: Rehm HJ, Reed G (eds) Biotechnology, vol 5, Food and feed production with micro-organisms. Verlag Chemie, Weinheim, Germany, p 1
19. Reed G, Nakodawithana TW (1991) Yeast technology. Van Nostrand Reinhold, NY, p 297
20. Stadhouders J, Jansen LA and Hup G (1969) Neth Milk Dairy J 23: 182
21. Foster EM (1962) J Dairy Sci 45: 1290
22. Robinson RK (1981) Dairy Ind Int 46: 15
23. Robinson RK (1983) In: Rehm HJ, Reed G and Dellweg H (eds) Biotechnology, vol 3, Biomass, Microorganisms for special applications, Microbial products I, Energy from renewable resources. Verlag Chemie, Weinheim, Germany, p 191
24. Philipp S (1984) Lebensmitteltechnik 10/84: 518
25. Gibson LF, Khoury JT (1986) Letters Appl Microbiol 3: 127
26. Lievense LC and Van 't Riet K (1992) In: Fiechter A (ed) Advances in Biochemical Engineering/Biotechnology, vol. 51, Springer, Berlin Heidelberg New York, p 65
27. Morichi T (1974) Jpn Agr Res Quart 8: 171
28. Kilara A, Shahani KM and Das NK (1976) Cultured Dairy Products J., May 1976: 8
29. Porubcan RS and Sellars RL (1979) In: Peppler HJ and Perlman D (eds) Microbial technology. Academic, London, p 59
30. Tsetkov TS and Brankova R (1983) Cryobiology 20: 318
31. Valdez GF, De Giori GS, De Riuz Holgodo AA and Oliver G (1983) Appl Env Microbiol 45: 302
32. Valdez GF, De Giori GS, De Riuz Holgodo AA and Oliver G (1983) Cryobiology 20: 560
33. Hill FF (1986) In: Chmiel H, Hammes WP and Bailey JE (eds) Biochemical engineering: a challenge for interdisciplinary cooperation. Int Congress Stuttgart Sept 1986. VCH Publishers, NY, p 199
34. Roser B (1991) Trends in Food Sci Technol 2: 166
35. Williams-Gardner A (1976) Industrial drying. George Godwin, London
36. Masters K (1985) Spray drying handbook. George Godwin, London
37. Filková I and Mujumdar AS (1987) In: Mujumdar AS (ed) Handbook of industrial drying. Marcel Dekker, NY, p 243
38. Hovmand S (1987) In: Mujumdar AS (ed) Handbook of industrial drying. Marcel Dekker, NY, p 165
39. Uhlemann H (1990) Chem-Ing-Tech 62: 822
40. Van't Land CM (1991) Industrial drying equipment: selection and application. Marcel Dekker, NY
41. Keey RB (1972) Drying: Principles and practice. Int series of monographs in chemical engineering, vol 13. Pergamon Press, Oxford
42. Lievense LC, Verbeek MAM, Meerdink G and Van 't Riet K (1990) Bioseparation 1: 161
43. Lievense LC, Verbeek MAM, Taekema T, Meerdink G and Van 't Riet K (1992) Chem Eng Sci 47: 87

44. Kuts PS and Tutova EG (1983) Drying Techn 2: 171
45. Nei T (1973) Cryobiology 10: 403
46. Orndorff GR and MacKenzie AP (1973) Cryobiology 10: 475
47. Brown AD and Melling J (1971) In: Hugo WB (ed) Inhibition and destruction of the microbial cell. Academic, London, p 1
48. Corry JEL (1973) Progr Ind Microbiol 12: 73
49. Daemen ALH (1981) Neth Milk Dairy J 35: 133
50. Zimmermann K (1987) Einflußparameter und mathematische Modellierung der schonenden Trocknung von Starterkulturen. Fortschr-Ber VDI (Reihe 14, no 36), VDI-Verlag, Düsseldorf, West Germany
51. Van der Lijn J (1976) Simulation of heat and mass transfer in spray drying. PhD Thesis, Agricultural University Wageningen, The Netherlands
52. Liou JK (1982) An approximate method for nonlinear diffusion applied to enzyme inactivation during drying. PhD Thesis, Agricultural University Wageningen
53. Lievense LC, Verbeek MAM, Meerdink G and Van 't Riet K (1990) Bioseparation 1: 149
54. Valdez GF, De Giori GS, De Riuz Holgodo AA and Oliver G (1985) Appl Env Microbiol 49: 413
55. Divies C, Lenzi P, Beaujeu J and Herault F (1990) Procede de preparation de micro-organismes inclus dans des gels sensiblement déshydrates, gels obtenus et leur utilisation pour la préparation de boissoins fermentées. French Patent 2633937. Champagne Moët & Chandon, France
56. Comings EW, Higa H, Myers JE, Koffler H and McLain HA (1977) Ind Eng Chem, Fundam 16: 12
57. Metwally MM, Abd El-Gawad IA, El-Nockrashy SA and Ahmed KE (1989) Egyptian J Dairy Sci 17: 35
58. Elizondo H and Labuza TP (1974) Biotechn Bioeng 16: 1245
59. Freeman RR, Tschernitz JL and Marshall WR (1964) Biotechn Bioeng 6: 473
60. Espina F and Packard VS (1979) J Food Prot 42: 149
61. Kim SS and Bhowmik SR (1990) J Food Science 55: 1008
62. Verhey JGP (1973) Neth Milk Dairy J 27: 3
63. McLain HA, Comings EW and Myers JE (1957) Chem Eng Prog 53: 282
64. Daemen ALH and Van der Stege HJ (1982) Neth Milk Dairy J 36: 211
65. Jung G (1988) Inocula of low water activity with improved resistance to temperature and rehydration and preparation there of USA Patent 4.755.468. Rhone-Poulenc SA, Paris, France
66. Bullock K and Lightbrown JW (1947) Q J Pharm Pharmacol 20: 312
67. Labuza TP, Roux JP, Fan TS and Tannenbaum SR (1970) Biotechn Bioeng 12: 135
68. Langejan A (1980) Active dried baker's yeast. USA Patent 4.217.420. Gist-brocades NV, Delft, The Netherlands
69. Langejan A and Khoudokormoff B (1982) Active dried baker's yeast. USA Patent 4.341.871. Gist-brocades NV, Delft, The Netherlands
70. Hill FF (1987) Alimenta 1987 No. 3: 73
71. Lehmann D (1984) ZFL 1984, No 2: 113
72. Clement P and Rossi J (1983) Preparation of dried baker's yeast. USA Patent 4.370.420. Societe Industrielle LeSaffre, Paris, France.
73. Pomper S (1989) Active dried yeast. USA Patent 4.797.365. Nabisco Brands Inc, Parsippany, NJ, USA
74. Clementi F and Rossi J (1984) Am J Enol Vitic 35: 181
75. Bera F, Renauld H, Rikir R and Thonart P (1988) Med Fac Landbouw Rijksuniv Gent 53: 2007
76. Taeymans D, Roelans E and Lenges J (1986) In: Le Maguer M and Jelen P (eds) Proc Int Congr Eng Food 4, Edmonton, Canada. Food engineering and process applications, vol 1. Elseviers, London, p 439
77. Roelans E and Taeymans D (1990) In: Spiess WEL and Schubert H (eds) Engineering and Food, vol 3, Advanced processes. Elseviers Applied Science, London, p 559

78. Hill FF (1985) Verfahren zum Herstellen eines trockenen lagerstabilen Bakterien-präparates. European Patent 0131114. Chemische Werke Hüls AG, Marl, Germany.
79. Zimmermann K and Bauer W (1990) In: Spiess WEL and Schubert H (eds) Engineering and Food, vol 2, Preservation processes and related techniques. Elseviers Applied Science, London, p 666
80. Prajapati JB, Shah RK and Dave JM (1987) Austr J Dairy Technol March/June 1987: 17
81. Kozachuk DN, Globa LI, Gordienko AS and Krasnobrizhii NY (1984) Microbiology: A translation of Mikrobiol 52: 336
82. Bagnoli E, Fuller FH, Johnson VJ and Norris RW (1973) In: Perry RH and Chilton CH (eds) Chemical Engineers' Handbook, 5th edition. McGraw-Hill, NY, p 12.2

An Overview on the Control System Design of Bioreactors

K. Shimizu
Department of Biochemical Engineering & Science,
Kyushu Institute of Technology, Iizuka, Fukuoka 820, Japan

The efficient control of bioreactor systems is becoming more and more important due to the recent significant development in biotechnology, computer science and knowledge engineering.

Here, the current status of the efficient control strategies for bioreactor systems are investigated together with the related areas of sensor technology, optimization, and state and parameter estimation.

Advances in Biochemical Engineering
Biotechnology, Vol. 50
Managing Editor: A. Fiechter
© Springer-Verlag Berlin Heidelberg 1993

1 Introduction

Recent development in biotechnology on microbial cultivation, recombinant DNA, hybridoma etc. has had a strong impact on bioindustry, and there is still increasing demand for further development for food, energy sources, medicine, and environmental treatment using microorganisms, plants, and animals.

The developments in genetic engineering have played a major role in creating the current excitement in biotechnology. Through these genetic manipulation techniques, a variety of products can be produced by using bacteria, yeast and animal cells as hosts. The other developments which have brought the current excitement may be biocatalysis such as the use of enzymes to produce useful substances and immunochemistry for the production of antibodies [1].

The biotechnology industry is evolving rapidly based on the above stated developments. It seems that it has already entered a new stage of growth and is about to leave behind the early stage of skepticism. The many biotechnology-based products such as pharmaceutical and health-care products, agricultural products, and chemicals have already been commercialized and attention is being focused on the transition from laboratory to marketplace [2]. Billions of dollars are being invested annually by many companies looking for capital appreciation and seeking a business opportunity.

Despite the steady progress in laboratory-scale research, however, there remain many problems associated with the scale-up of bioprocesses. Since most biochemical processes create very dilute and impure products, there is a great need to increase volumetric productivity and to increase the product concentration.

As a result, the scale-up of bioprocesses requires a large investment in the development of an efficient processing technology. In this regard, significant work is needed to optimize the design and operation of bioreactors to make production more efficient and more economical.

Historically, it has been believed that the increased productivity can be attained through strain and media improvements. Due to the recent significant efforts, however, potential benefits have been recognized for the application of advanced control techniques. The ability to control bioprocesses at their optimal states accurately and automatically is now of considerable interest to many bioindustries since it can enable them to reduce their production costs and increase the yield while at the same time maintaining the quality of metabolic products.

It should be noted, however, that the control system design of bioreactors is not straightforward due to:

1) the lack of accurate mathematical models which can describe the cell growth and metabolite production,
2) the time-varying and nonlinear nature of the systems for batch and fed-batch operation,

3) the lack of reliable on-line sensors which can detect the important state variables, and
4) the slow responses of the process in particular for cell and metabolite concentrations.

To cope with the first and second problems, the control system must be robust for model uncertainties with the ability to reject disturbances. As for the third problem, the effective on-line sensors are limited for use in many cases so that some sort of observer and/or parameter estimator needs to be designed from the available measurement variables. As for the fourth problem, the slow response is significant in particular in mammalian cell culture, and the predictive type of control may be considered in such a case.

Many control strategies have been proposed so far and applied in practice to overcome the above problems, and many others are still in progress [3].

In the following, starting with the brief review of the particularly related areas such as "On-Line Sensor Developments", "Optimization" and "State and/or Parameter Estimation", the efficient control strategies are reviewed with their application to bioprocesses including our recent contribution.

2 On-Line Sensor Developments

The activity of a particular enzyme is controlled by the concentration of several medium components. Well-known are the enzyme regulations by carbon compounds (e.g., glucose), nitrogen compounds (e.g., ammonia), phosphates as well as induction of enzymes by their substrates.

In order to achieve the full biological potential of the cells, the environmental conditions must be maintained at the optimum. The first step towards this goal is the identification of the key components, which requires on-line analysis of the medium and cellular components. This information is critical for success in bioprocess control [4]. However, in situ techniques are restricted, at present in many cases, to the measurement of DO (dissolved oxygen) concentration, DCO_2 (dissolved CO_2) concentration, pH, temperature, and redox potential [5].

The main problems for on-line monitoring may be the on-line sample conditioning, calibration, and blank measurements. The flow injection analysis (FIA) uses a carrier flow with a reagent in which a small amount of sample pulse is injected. This causes a large dilution of the cultivation medium and reduces the probability for cell growth in the analyzer. The analyte forms a product with the reagent, and the concentration peak height of which is measured in the detector. This method is becoming popular since it requires only a small amount of reagent and sample, and yet has short reaction times, and high flexibility.

The concentration of microorganisms and animal cells are important variables. The most common methods to determine the cell (biomass) concentration in situ in cultures of microorganisms are the measurement of the transmitted light (turbidity) [6] and scattered light (nephelometry) [7]. These methods, however, are not selective enough. They cannot distinguish what are solid particles, dead, or viable cells.

NADH and NADPH in the viable cells can be excited by UV light, and the induced fluorescence can be measured by suitable detectors. This NAD(P)H fluorescence has a large potential as a tool for monitoring, studying and controlling cultivations [8–11]. In some special cases, there is a linear relationship between the cell mass concentration and the fluorescence intensity (e.g., *Zymomonas mobilis* [12], *Methylomonas mucosa* [13] and *Pseudomonas putida* [14]).

The carbon source is one of the most important medium components, and the glucose sensor is quite important for various cultures. The fundamental characteristics of an automatic glucose analyzer were examined [15]. The glucose sensor is a dual cathode type which has an immobilized glucose oxidase membrane coupled with an oxygen sensor. Using this sensor combined with an automatic sampling device, the glucose concentration was kept low for the fed-batch culture of *Saccharomyces cerevisiae* and *Micrococcus ruteus*.

Recent development in on-line techniques covers various items such as on-line aseptic sampling, on-line continuous flow analysis, on-line flow injection analysis, on-line HPLC (high performance liquid chromatography), and on-line MS (mass spectrometry). HPLC is a standard technique in the chemical industry [16]. However, it is rarely applied for on-line process monitoring in biotechnology. When using very pure chemicals and degassed solvents, stable and long-range operation is possible [17].

Inexpensive MS is becoming popular for monitoring exhaust gas composition of cultivations. Recently, dissolved gas and volatile components have also been analyzed on-line [18].

Fully automated analyzer systems allow the real-time elemental balancing and evaluating nonmeasurable process variables [19]. The on-line measurement of the concentrations of all precursors and key medium components in bioreactors allows the investigation of the in vivo biosynthesis of secondary metabolites during their production as a function of the medium composition [20] as well as the evaluation of the dynamic relationship between cell regulation and reactor control. These are prerequisites for the development of a realistic dynamic model for advanced process control.

3 Optimization

In the last two decades, a number of optimal operation strategies have been reported for the production of metabolic products. Although the variety of metabolic products, the complexity of bioprocesses, a number of control variables and the fuzzy nature of the system make the problem difficult to solve, the development of optimal operation strategies definitely contributes to the increase in the overall profits. The optimal operation of various bioprocesses has been reviewed by several researchers [21, 22].

Note that the development of the sound mathematical model that adequately describes the dynamic behavior of the system is vital to the success of optimization. Many mathematical models have therefore been proposed in the past to describe substrate utilization, growth of microorganism, and product formation. These may be broadly classified into two groups: structured models and unstructured models. Although the former can provide greater insight into the complicated mechanisms in growth metabolism than the latter, they contain many parameters and become unwieldy for practical use in particular from an engineering point of view. It should be noted that the segregated models may be useful for the stability analysis of plasmids. On the other hand, the advantage of using the unstructure models lies in their simplicity and the convenience in utilizing specific sets of data. It should be kept in mind, however, that the model uncertainty must always be taken into account for the optimization and control in this case.

For the efficient bioreactor operation, several operation modes such as batch, fed-batch, their cyclic repetition and chemostat have been considered, and some of them have been extensively employed in industry.

Fan and Wang [23] considered an enzymically catalyzed reaction occurring in a series of stirred tank reactors. The optimization was made using the discrete type of maximum principle. The objective function to be maximized was the concentration of the product at the last stage. The optimal temperature or pH could thus be obtained. Wen et al. [24] extended the work of Fan and Wang [23] to the case where a portion of the output from the last stage was recycled to the first stage.

Blanch and Rogers [25] developed a mathematical model which describes the production of gramisidin S from *Bacillus brevis* based on the idea of cell states, "immature" and "mature". They assumed that the aging rate from immature cell to the mature cell was constant and that the mature cell was divided into two immature cells by binary fission. They also applied the discrete maximum principle to find the optimal operation in a multistage bioreactor system using the model developed [26]. The objective function was formed taking into account several factors such as the selling price of the product, the cost of vessels, the cost of the substrate in the upstream, and the cost of separation (extraction) in the downstream. The optimization was made with respect to the residence time, pH, and temperature in each reactor.

Alcohol production is one of the oldest processes. Although the alcohol production has been dominantly used for producing beverages, the recent rising cost of petroleum has increased further interest in the production of bioalcohol. Although many systems have been operated as batch processes, recent trends are the development of the continuous systems which reduce production costs through ease of operation by eliminating reactor downtime between batches, and have the additional advantage of being generally simpler to control. It must, however, be noted that continuous operation suffers from the risk of contamination and genetic instability by mutation. Many processes for alcohol production have been developed so far to increase productivity. The detailed review on this topic was made by Maiorella et al. [27], and is not duplicated here. It should, however, be added that Lee et al. [28] studied the multistage reactors with cell recycle by computer simulation, and Ciftci et al. [29] considered the cell feeding strategy for multistage alcohol production.

Ohno et al. [30] considered the problem of determining the optimal operation patterns such as batch, fed-batch, and continuous for the maximization of the production of a metabolite. The problem can be solved by the application of the maximum principle. However, the following situation often occurs, namely the feed flow rate, one of the control variables, becomes identically zero over some finite time interval. This is the so-called "singular" control the law of which is difficult to derive in general. One way to overcome this difficulty is to utilize the generalized Legendre-Clebsh condition [31] but the required computation tends to be complex. Ohno et al. [32] considered transforming the problem into the nonsingular problem by the application of Kelley's transformation [33]. The optimal operation patterns were thus obtained for lysine production (Ohno et al. [30]).

Constantinides et al. [34] developed a mathematical model for penicillin production based on the experimental data obtained from the production of *P. chrysogenum* performed at different temperatures. The cell growth was modeled by the logistic equation, and the penicillin production was modeled by the assumption that only the mature cells can produce penicillin. The model parameters were temperature-dependent, and the optimal temperature policy was obtained by the application of the maximum principle [35]. Rai and Constantinides [36] studied the optimal temperature and pH policies for the gluconic acid production by the maximum principle.

Fishman and Biryukov [37] considered the problem of optimal substrate feeding policy for penicillin production. They incorporated the concept of cell age of the culture and its effect on penicillin production into the kinetic model. They found that the optimal feeding mode is a bang-singular-bang profile. It was stated that the improvement of about 7% in the final penicillin concentration was attained using such optimal feeding policy instead of using the constant feeding policy. Ohno et al. [32] also considered the optimal feeding policy for amino acid production and solved it by the application of Green's theorem.

The optimization of fed-batch cultivation attracted many researchers in the past, and the significant efforts have also been made recently [38–41].

The repeated batch or fed-batch cultivation was investigated in the past by several researchers for the improvement of cell and metabolite production. Such a culture may be considered as an intermediate culture between batch or fed-batch and continuous cultures. Pirt [42] may have been the first to discuss the repeated fed-batch cultivation and consider the quasi-steady state of such cultivation. Later, Keller and Dunn [43] studied the effect of the initial volume and the feed flow rate on the biomass productivity of cyclically operated variable-volume reactors by computer simulation. Dunn et al. [44] considered the possibility of calculating the biomass and metabolite productivities for cyclic operation graphically. They concluded that although a chemostat would always have higher productivity, a cyclic fed-batch might for certain kinetic forms have a higher productivity than a cyclic batch system. Weigand [31] determined the optimal operation of repeated fed-batch operation using the maximum principle for the case of constant yield. Mori et al. [45] compared the cell productivities for three kinds of culture system experimentally using an ethanol-assimilating yeast, *Candida brasicae* and showed that they increased in the order of fed-batch, continuous, and repeated fed-batch cultures.

The development of a microbial process for the formation of a metabolic product is aimed at maximizing several items such as the yield of product per g of substrate, the concentration of the product, the rate of product formation etc. Although many researchers have restricted their analyses to one of the above criteria, we have pointed out the importance of using a vector-valued objective function taking into account some of the above items [46–52]. It should be noted that the comparison of the performance of the operation modes was based on the noninferior set [53].

Finally, it may be useful to mention the start-up and the optimal regulation problem for chemostat operation. D'Ans et al. [54, 55] considered the optimal regulation problem where the problem was to find the optimal control policy which quickly returns the process to its optimum point. They showed that it could be accomplished by using the Green's theorem to predict the optimal trajectory of the dilution rate which would also maximize the biomass production. An algorithm for the similar optimal control problem was described by Muzychenko et al. [56]. Yamane et al. [57] applied the same technique to find a time-optimal strategy for the start-up of a chemostat. The start-up strategy of chemostat is of practical interest in large-scale industrial production of microbial cells for single cell protein (SCP), and baker's yeast, because a constant quality of the cell mass should be attained as soon as possible, and because very long term continuous culture is seldom possible due mainly to contamination and genetic change. Takamatsu et al. [58] studied the problem of optimal operation for the continuous production of amino acid. They developed a mathematical model based on the pilot plant data for solving two problems such as (1) the time-optimal operation which moves the system from a given initial state to the desired steady state, and (2) the optimal operation to maximize the production of amino acid within the given operation time. They combined the maximum principle and the Green's theorem to arrive at the optimal operation strategy.

4 State and Parameter Estimation

The use of optimal or suboptimal estimation techniques to reconstruct the unknown system states has become increasingly important in particular for bioprocess control since the number of available on-line sensors is limited in many cases in practice. Although the Kalman filter developed by Kalman [59], and Kalman and Bucy [60] is of interest, it was assumed that the dynamics of the system be expressed by the linear system equations. The dynamic behavior of most of the biological processes is, however, nonlinear in nature.

The extended Kalman filter which can be applied to the nonlinear system is, therefore, of significant interest from the practical application point of view [61–64]. In particular, the on-line estimation of the specific growth rate is of practical interest [65, 66].

For the estimation of the specific growth rate, Mou and Cooney [67] considered the use of the mass balance in the penicillin process. They controlled the specific growth rate at a high level in the early stage of the culture to increase the cell mass, and then kept it low in the late stage of the culture to promote the penicillin production. Several other methods of estimating the specific growth rate have been developed by several researchers [68, 69] without using the extended Kalman filter. Although the extended Kalman filter has been applied in practice and showed some success, its direct application sometimes gives unsatisfactory results. This is due to the fact that the extended Kalman filter was derived by the application of the Kalman filter algorithm to the linearized nonlinear system. It is, therefore, expected that the filter performance can be improved by eliminating several approximations associated with linearization [70]. Several iterations must, however, be incorporated in the calculation in this case.

It should be noted that the success of the Kalman filter depends largely on the accuracy of the process model. Noting that most bioprocesses are subject to significant model uncertainties with time-varying nature, it is quite natural to consider parameter estimation as well as state estimation. The joint state and parameter estimation is, therefore, the current research topic in this area [71].

It should be mentioned that the parameter identification and state estimation should be considered taking into account the nonlinearity of the process model [72–74].

5 Efficient Control Strategies

The use of computer control has attracted great attention recently since many bioprocesses prevent us from applying the conventional PID controller. Aiba [75] and Constantinides [21] reviewed the process control and optimization, and Cooney [76] addressed some perspectives of computer application.

As early as 1962, Yamashita et al. [77] examined DDC (direct digital control) in glutamic acid production. A pilot-scale reactor was equipped with controllers for temperature, vessel pressure, pH, air flow rate, and foam. Analyses were made for outlet gas composition, glutamate concentration in the medium, DO (dissolved oxygen) concentration, and microbial population density.

DDC was utilized in the batch production of penicillin in 1969 [78]. In 1971, the idea of data acquisition, analysis, and computer control of processes was presented [79]. In batch or fed-batch culture, the model parameters significantly change in the course of the process. The computer is an ideal tool for the treatment of such problems because of its real-time capabilities [80].

A computerized bioreactor system having complete feedback capabilities for use in a research environment has been described by Mohler et al. [81]. The features of this system include the flexibility to handle various modes of reactor operation for different organisms, the ability to monitor and control the experimental conditions efficiently and accurately, and ease of operation. Several other applications of computer control may be found elsewhere [82–87]. Recent computer control techniques may be found in several books [88, 89].

A survey has been presented for the control of the fed-batch modus [90]. In the following, a survey is presented for the variety of control strategies applied to many bioprocesses.

5.1 Gain Scheduling

Due to the inherent time-varying characteristics of a batch or fed-batch bioreactor, proper tuning of the PID controller is a very demanding task. For example, the ultimate gain for the feedback control of DOT (dissolved oxygen tension) increases considerably as time proceeds or the OUR (oxygen uptake rate) increases in accordance with the increase in the cell concentration [91]. This means that a controller properly tuned for a low OUR will produce a sluggish response with significant offset as time proceeds, and conversely that a controller tuned for a high OUR may cause instability at the initial state of the process. This clearly indicates the need for changing the controller gain as time proceeds or as a function of OUR to maintain good control performance.

A similar situation may be found in many cases such as the feedback control of ethanol concentration in baker's yeast cultivation [87] to prevent ethanol formation which occurs in response to the Crabtree effect.

5.2 Programmed Control with Feedback Compensation

The next approach is based on the idea that some of the growth parameters may be roughly obtained from the past batch data. Then, for example, the expected feed rate can be computed as a function of time so that the substrate concentration in the reactor is kept at the desired level. Takamatsu et al. [92] showed the usefulness of this control scheme by computer simulation for a baker's yeast fedbatch culture with several types of feedback compensators. The measurement variable was either the cell concentration or the specific growth rate. Shimizu et al. [93] verified experimentally the efficiency of the algorithm by estimating the specific growth rate.

While their experimental results were obtained in the range of low levels of cell concentration, we carried out the experiment from a low level to a high level of cell concentration by measuring the glucose concentration in the reactor on-line with the P or PI controller as a feedback compensator [94]. The control quality seems to be satisfactory, while it tends to give an offset at around the end of the exponential growth phase.

5.3 Optimal Control with Model Identification

Based on the assumption that the process model is not known a priori but the time-series data of input and output variables are available on-line, it may be reasonable to assume linear model and identify model parameters by techniques such as (recursive) least squares method etc. Then the input to the process can be determined so that optimal control is attained in the sense that the squared error between the set point and the output variable is minimized.

Wu et al. [95] employed this approach for the fed-batch culture of baker's yeast. The output variable was the specific growth rate obtained by the measurement of OUR and the cell concentration estimated from the balance equation of oxygen. It was shown experimentally to be effective to keep the specific growth rate at the appropriate vales by adjusting the nutrient feed rate to prevent undesirable ethanol production. We also made the experiment with the control system mentioned above [94]. The glucose concentration was measured on-line and was controlled at some specified value. The control quality seems to be satisfactory except at the final stage of cultivation where the output variable tends to fluctuate.

Williams et al. [96] considered the multivariable type of optimal control with the measurement variables of RQ (respiratory quotient) and DO concentration. The manipulated variables were sugar feed rate and agitation rate.

5.4 Adaptive Control

Dochain and Bastin [74] studied the application of nonlinear adaptive control to a microbial growth system. They proposed an adaptive minimum variance control algorithm for two different types of problems: substrate concentration control and production rate control, and they showed its effectiveness by computer simulation.

As stated in Sect. 5.1, proper tuning of controller parameters is quite important for batch or fed-batch cultivation. Radjai et al. [97] applied a simple automatic self-tuning method to the control of redox potential by manipulating the agitation speed in amino acid production. The method is the one proposed by Astrom. The "ultimate gain" and the "ultimate period" were obtained on-line based on the response of the system under relay control and the PI parameters were determined based on the Ziegler-Nicols design criteria for parameter settings. It was shown that the total amino acid yields could be significantly increased with this control system as compared with the ones reported in the literature.

Dekkers and Voetter [98] applied the self-tuning LQG control strategy for a fed-batch baker's yeast cultivation. Noting that OUR is governed by the slow mode which is growth-associated and that the dynamics of RQ is governed by the fast mode which is associated with the rapid uptake of the supplied glucose, the glucose feed rate was manipulated based on the measured OUR with the self-tuning control for the specified RQ-value. The experimental result is quite satisfactory for the control of RQ except when an estimator wind-up failed after about 1 h.

Verbruggen et al. [99] also considered the self-tuning control based on the pole-assignment principle, and applied it to the fed-batch baker's yeast process. The control system is a cascade configuration (multiloop control) in which the self-tuning controller compensates for the RQ in the inner loop while an integrating controller compensates for the CER (carbon dioxide evolution rate) of which set point must vary in accordance with the value of OUR in the outer loop.

Frueh et al. [100] also applied a self-tuning control strategy to penicillin V where *Penicillium chrysogenum* was cultivated with suspended pellets in an air-lift reactor having an outer loop. Because of the significant change of the system parameters, efforts to control the DO concentration by means of the aeration rate with a conventional PI-controller failed. Then the self-tuning controller was employed to control the DO concentration. To avoid offsets, the minimum variance controller was multiplied by a proportional integral action term. The self-tuning controller was successfully applied to control the DO concentration.

Another application of a self-tuning controller was made to control the DO concentration in an activated sludge pilot plant [101]. Since wastewater aeration systems are the most energy demanding unit processes at municipal sewage treatment plants, significant efforts have been directed toward maintaining the DO concentration at a minimum set point level. Yust and Howell [101] achieved minimum variance control when controlling the DO concentration by real-time computer manipulation of the air flow rate.

5.5 Predictive Control

The major problem in the substrate addition/biomass control in the penicillin process is that the applications of steps, pulses or PRBS (pseudo random binary sequence) signals to the feed can subject the mould to conditions which are unfavorable to growth. Montague et al. [102] applied the GPC (generalized predictive controller), a long-range receding-horizon predictive-type algorithm, developed by Clark and Mohtadi [103] to penicillin production. Biomass was estimated using an extended Kalman filter from on-line measurements of CPR and culture volume. The GPC was applied to control the estimated biomass to follow the predefined trajectories which were obtained in the industrial plant. The application of this control technique offers the possibility of tight biomass trajectory control. In particular, the GPC has been shown to be effective in control with its inherent low overshoot property, ease of tuning, and ability to include pre-programmed future set points.

The DMC (dynamic matrix control) is an algorithm developed by the Shell Oil Company in Houston [104] and is in the category of predictive control. It has been demonstrated that the DMC algorithm is particularly well suited to the system with unusual dynamics such as slow response and inverse response.

On the other hand, it has been desired to locate a competitive mixed-culture equilibrium point and establish steady-state operation at that point for any selected ratio of the two microbial species. For the competitive mixed culture, a small perturbation in a culture variable such as pH or dilution rate away from its metastable equilibrium state does not lead to a new steady state but rather to the eventual washout of one of the two microorganisms. Goochee et al. [105] applied the DMC algorithm for the control of the competitive mixed culture of the yeast *Candida utilis* and the bacterium *Escherichia coli*. Noting that the dynamic response is slow, and that the system is nonlinear with model uncertainty, they modified the DMC algorithm by introducing new tuning parameters such as the desired rate of approach to the final set point and the projection vector rotation factor. The experimental result shows some success off where the output variable was OD (optical density), which depended on the proportions of bacteria and yeast due to the difference in yields of the two microorganisms, and the manipulated variable was pH.

5.6 Nonlinear Control

To overcome the nonlinear problem which limits the application of the well-established linear control theory, it may be useful to consider some transformation which converts the original nonlinear system into linear system. Global linearization refers to a set of methods for constructing algebraic transformations of the input and state variables of a process model such that the input-output behavior mimics a linear system [106]. Hoo and Kantor [107] demonstrated the applicability of a nonlinear multivariable controller for a mixed-culture bioreactor system by computer simulation. The reactor studied is a chemostat of two species of the same strain which compete for a single rate-limiting substrate. The growth of one species without having plasmids is inhibited by the addition of an external agent such as an antibiotic, while the growth of the second species having plasmids renders the external agent inactive. The stability and the control problems for the continuous culture of plasmid-bearing cells are the current major topics in recombinant-DNA studies [108]. It was shown [107] that the stable closed-loop operation with coexistence of the cell populations was possible by the linearizing transformation which could be attained by choosing the dilution rate and the total inhibitor addition rate as control variables.

Recently, Henson and Seborg [109] studied the nonlinear control strategies for continuous cultivation by computer simulation. Assuming that the control objective is to regulate the reactor near the optimal productivity, they showed that exact input-output linearizing control employing the dilution rate as the manipulated input provides excellent regulatory behavior, and that if the feed substrate concentration is chosen as the manipulated variable, the nonlinear control is problematic.

5.7 Repetitive Learning Control

Given a desired output trajectory which may be obtained by the past operational experiences, certain kinds of iterative learning control can be effective in that the process output converges to the desired trajectory as the operation is repeated. In this control strategy, the input to the process at the present operation is determined taking into account the input/output data of the past operations to achieve the desired set point tracking. We applied several repetitive control strategies to baker's yeast cultivation, ethanol and penicillin production, and studied the possibility of introducing such control schemes by computer simulation [110]. It was found that some compensator is necessary to cope with the environmental changes and the disturbances imposed in the current operation. Shimizu et al. [111] applied this strategy for the control of pH in the esterolysis reaction of the n-acetyltyrosine ethyl ester by chymotrypsin, and showed some experimental success.

5.8 On-Line Optimizing Control for (Fed-)Batch Culture

So far, we have assumed that the desired output trajectory was given a priori. Then the control effort was to track the predetermined output trajectory either by one cultivation or by repetitive cultivations. The important thing in practice is, however, to determine the output trajectory based on some economic criterion. Significant efforts have, therefore, been focused on this problem using some mathematical models. The optimal output trajectories were obtained by such techniques as the maximum principle, Miele's extremization method. Although the results obtained by such methods may be of some help in getting insight into the optimization, the output trajectories so obtained are by no means optimal since the mathematical models are more or less subject to uncertainty. It is, therefore, necessary to introduce the idea of on-line optimizing control to overcome the problem.

Kishimoto et al. [112] sought the optimal feeding rate on-line by utilizing the past cultivation data. The optimization was made with the aid of a dynamic programing technique and the application was made to glutamic acid production by *B. divaricatum* with some success.

5.9 On-Line Optimizing Control for Continuous Culture

The objective of on-line optimizing control for continuous cultivation is to keep tracking the optimal state in accordance with enzymatic deactivation and/or environmental changes. Although a wide variety of techniques have been published, the method of on-line optimization by dynamic model identification [113] is particularly attractive for bioreactor systems the dynamics of which are significantly slow. Several applications have been published showing some success [114–116].

We also proposed several on-line optimizing control strategies and tested them by computer simulation [117] and by experiment [118]. The control task was divided into two, one of which was to search for the optimal operating point and pass the set point to the lower layer of which task was to make the process output follow the set point as quickly as possible. It was shown to be effective for the upper layer to express the objective function as a polynomial with respect to the measurement variable and this can be used to find the optimum point. It was also shown to be quite effective to use the discrete type of self tuning PID controller and the optimal controller compensated for the interaction between the control loops in the lower layer. Application was made to the cell recycle system for the production of lactic acid and baker's yeast cultivation.

5.10 Control for Different Time Scales

In a continuous immobilized cell column reactor without bead removal for penicillin-G production, the growing cells remain in the reactor, which means that the system cannot reach steady state. Although the biomass concentration increases exponentially at a constant growth rate, the concentrations of substrate, penicillin-G, and precursor can be kept constant. This state may be defined as a QSS (quasi-steady-state). In such a situation, it was shown [119] that (a) growth rate dynamics are very fast, (b) penicillin concentration dynamics are very slow, (c) penicillin concentration transients induce growth rate transients, (d) growth rate transients do not induce any penicillin concentration transients. With this in mind, Kalogerakis et al. [119] considered a simple control strategy where the PI controller was used to manipulate the dilution rate in speeding up the penicillin concentration transients. For the control of the specific growth rate, they considered a steady-state controlled by manipulating the feed glucose concentration taking into account the interaction between the two control loops. A similar situation can be seen in fedbatch cultivation. Boyle

[120] noted the different time scales for the substrate concentration transients and the biomass concentration transients in baker's yeast cultivation.

5.11 Knowledge-Based Control

Although it is fairly difficult to describe exactly the behavior of microorganisms by means of mathematical expressions in many cases, it may be possible to make use of the information obtainable from operators' intuition and experience for the control of bioreactors. The expert system characterizes the process dynamics by symbolic and logic description to check data consistency, equipment failure, and contamination for efficient operation [121, 122].

On the other hand, fuzzy sets theory has been paid a great deal of attention recently and has been applied to various bioprocesses such as glutamic acid [123], antibiotic [124, 125], SCP- [126], and coenzyme Q_{10} production [127], plus sake brewing [128, 129]. Since the process state significantly changes as time proceeds in the batch or fed-batch type of operation, the membership functions should be changed in accordance with the process state [126, 130]. We proposed a neuro-fuzzy control where the membership functions were changed in accordance with the changing patterns of the state variables where the patterns were recognized on-line by neural networks [131].

Very recently, Konstantinov and Yoshida [132] reviewed the knowledg-based control strategies and summarized their functions as follows:

1) Input data validation.
2) Identification of the state of the cell culture.
3) Detection and diagnostics of instrumentation faults.
4) Supervision of conventional control.
5) Communication with the user.
6) Plantwide supervision and scheduling.

It should be noted that the control system must eventually be extended to the related upstream or downstream processes.

Konstantinov and Yoshida [126] proposed a methodology for the control of bioprocesses based on the expert identification of the physiological state of the cell population. The physiological state was defined quantitatively by a set of specially selected variables that form the physiological state space of the culture. Upon transfer from one to another state, it often exhibits variable structure behavior, and therefore different control strategy must be applied.

6 Conclusions

With the recent significant development in biotechnology, it is increasingly important to control bioreactor systems. With an overview given in this article, it can be seen that the variety of control strategies have been applied to many bioprocesses. A table of pros and cons of the different techniques and possible fields of application would give the inexperienced reader an idea of the commonly used approach and orientation to what they can afford. In the following, some evaluation is made for different approaches in an attempt to draw some conclusions.

In batch or fed-batch cultivation, the (local) dynamic behavior significantly changes as the cell concentration increases. The idea of "gain scheduling" is to change the proportional gain of the PID controller in accordance with the cell concentration. It should be noted that in this approach, the ultimate gain must be predetermined as a function of cell concentration or OUR. Since the dynamic behavior will differ from cultivation to cultivation, the robustness consideration is critical. Otherwise, the applicable range is limited.

The "programmed control" strategy is based on the mathematical models developed by the mass and/or heat balances. Since the model parameters are the lumped-approximation, some feedback compensation is required. This feedback compensator is usually designed based on the linear control theory. In general, the model-plant mismatch becomes significant as time proceeds. Therefore, the success of this approach depends on the consideration of robustness (closed-loop stability under model uncertainty).

The "self-tuning" type of adaptive control strategy does not require the mathematical model. The linear model is identified on-line based on the time-series data of input and output variables. Although this method and also the "model-based adaptive control" show wide applicability, the control system design is not straightforward.

Although the "nonlinear control" strategy can cope with the nonlinearity associated with bioprocesses, this method relies heavily on the accuracy of the mathematical models. The theoretical part has been well investigated with some simulation studies, but the practical application is quite limited so far.

The idea of "repetitive learning control" strategy is of interest utilizing the past experimental data. This approach has been successfully applied to the control of robot manipulators. The direct application of this method to bioprocesses, however, may not lead to success since the dynamic behavior will change from cultivation to cultivation due to enzymatic deactivation and unpredictable disturbances in the culture environment. The approach may become promising by combining this technique with knowledg-based control strategies which can cope with vague and uncertain phenomena.

From the practical application point of view, the "optimizing control" may be quite important. On-line optimizing control for continuous cultivation has long been a subject in the chemical industry. Wealth of knowledge and experiences accumulated for the control of chemical plants such as distillation systems can be applied to bioprocesses with some modification for the on-line model identification. Now, the problem is the on-line optimizing control for batch or fed-batch type of cultivation since such modes of operations are employed exclusively in industry. There may be several approaches to attain optimizing control for such modes of operations. One way is to consider the hierarchical control structure where the task of the upper layer is to recognize or learn the dynamics of the whole stage of cultivation based on the past cultivations, and to find the optimal trajectory. This may be carried out using artificial neural networks [133]. The main task of the lower layer is to track the optimal trajectory determined in the upper layer, but the problem is not so simple since the current process state may not be the same as that corresponding to the optimal trajectory. Then the optimal trajectory must be modified taking into account the current state condition. One idea to cope with this problem is to employ the predictive type of control strategy [134].

It should be noted that the important thing in considering the control problem is not simply to apply sophisticated control theory but to consider it with deep understanding of the dynamics of the physiological state changes. It is, therefore, quite important to investigate how the physiological state changes in relation to genetic change and environmental change, and this so-called "metabolic engineering" [135, 136] is becoming quite promising from the control point of view.

7 References

1. Wang DIC (1988) Biotechnology: Status and Perspective. AIChE Monograph Series. 84:1
2. Shamel RE, Chow JJ (1987) Chem Eng Prog 83:41
3. Fish NM, Fox RI, Thornhill NF (eds) (1989) Computer applications in fermentation technology: Modelling and control of biotechnological processes. Elsevier Applied Science, London
4. Schügerl K (1991) On-line analysis of broth. In: Rehm HJ, Reed J (eds.) Biotechnology. VCH. Weinheim, p 149
5. Schügerl K (1991) Common instruments for process analysis and control. In: Rehm HJ and Reed (eds.) Biotechnology. VCH. Weinheim, p 5
6. Iijima S, Yamashita S, Matsunaga K, Miura H, Morikawa M, Shimizu K, Matsubara M, Kobayashi T (1987) J. Chem. Tech. Biotechnol. 40:203
7. Readon KF, Scheper TH (1991) Determination of cell concentration and characterization of cells. In: Rehm HJ and Reed J (eds.) Biotechnology. VCH. Weinheim, p 179
8. Ristroph DL, Watteeuw CM, Arminger WB, Humphrey AE (1977) J. Ferment. Technol. 55:599
9. Beyeler W, Einsele A, Fiechter A (1981) J. Appl. Microbiol. Biotechnol. 13:10
10. Scheper T, Schügerl K (1986) Appl. Microbiol. Biotechnol. 23:440
11. Zabriskie DW, Humphrey AE (1978) AIChE J. 24:138
12. Scheper T, Lorenz T, Schmidt W, Schügerl K (1987) Annals of the NY Acsd. Sci. 506:431
13. Doran PM, Bailey JE (1987) Biotechnol. Bioeng. 29:892
14. Boyer PM, Humphrey AE (1988) Biotechnol. Letters 2:193
15. Mizutani S, Iijima S, Morikawa M, Shimizu K, Matsubara M, Ogawa Y, Izumi R, Matsumoto K, Kobayashi T (1987) J. Ferment. Technol. 65:325
16. Horvath C (ed.) High performance liquid chromatography, Advances and perspective, Academic, New York
17. Moller J, Hiddessen R, Niehoff J, Schügerl K (1986) Anal. Chim. Acta 190:195
18. Heinzle E, Dunn IJ (1991) Methods and Instruments in Fermentation Gas Analysis. In: Rehm HJ and Reed G (eds.) Biotechnology. VCH. Weinheim p 27
19. Bellgardt KH, Kuhlmann W, Meyer HD, Schügerl K, Thoma M (1986) IEE Proc. 133:226
20. Holzhauer-Rieger K, Zhouw, Schügerl K (1990) J. of Chromatography 499:609

21. Constantinides A (1979) Ann NY Acad. Sci.: 193
22. Yamane T, Shimizu S (1984) Adv. Biochem. Eng./Biotech. 30:147
23. Fan LT, Wang CG (1963) Biotech. Bioeng. 5:201
24. Wen CY, Chang TM, Fan LT, Ko YC, Knieper, Jr PJ (1967) Biotech. Bioeng. 9:113
25. Blanch HW, Rogers PL (1971) Biotech. Bioeng. 13:843
26. Blanch HW, Rogers PL (1972) Biotech. Bioeng. 14:151
27. Maiorella B, Wilke CR, Blanch HW (1981) Adv. Biochem. Eng. 20:44
28. Lee JM, Pollard JF, Coulman GA (1983) Biotech. Bioeng. 25:497
29. Ciftci T, Constantinides A, Wang SS (1983) Biotech. Bioeng. 25:2007
30. Ohno H, Nakanishi E, Takanatsu T (1978) Biotech. Bioeng. 20:625
31. Weigand WA (1981) Biotech. Bioeng. 23:249
32. Ohno H, Nakanishi E, Takamatsu T (1976) Biotech. Bioeng. 18:847
33. Kelley JH (1965) J. SIAM Control 2:234
34. Constantinides A, Spencer JJ, Gaden Jr EL (1970) Biotech. Bioeng. 12:803
35. Constantinides A, Spencer JJ, Gaden Jr EL (1970) Biotech. Bioeng. 12:1081
36. Rai VR, Constantinides A (1973) AIChE Symp. Ser. 132(64):114.
37. Fishman VM, Biryukov VV (1974) Biotech. Bioeng. Symp No. 4:647
38. Lim HC, Tayeb YJ, Modak JM, Bonte P (1986) Biotech. Bioeng. 28:1408
39. Modak JM, Lim HC, Tayeb YJ (1986) Biotech. Bioeng. 28:1396
40. Modak JM, Lim HC, (1987) Biotech. Bioeng. 30:528
41. Park S, Ramirez WF (1988) AIChE J. 34:1550
42. Pirt SJ (1974) J. Appl. Chem. Biotech. 24:415
43. Keller R, Dunn IJ (1978) J. Appl. Chem. Biotech. 28:784
44. Dunn IJ, Shioya S, Keller R (1979) Ann NY Acad. Sci. 326:127
45. Mori H, Yamane T, Kobayashi T, Shimizu S (1983) J. Ferment. Technol. 61:391
46. Matsubara M, Hasegawa S, Shimizu K (1985) Biotech. Bioeng. 27:1214
47. Hasegawa S, Shimizu K, Kobayashi T, Matsubara M (1985) J. Chem. Tech. Biotech. 35B:33
48. Shimizu K, Kobayashi T, Nagara A, Matsubara M (1985) Biotech. Bioeng. 27:743
49. Hasegawa S, Matsubara M, Shimizu K (1987) Biotech. Bioeng. 30:703
50. Honda H, Mano T, Taya M, Shimizu K, Matsubara M, Kobayashi T (1987) Chem. Eng. Sci. 42:493
51. Shi Z, Shimizu K, Iijima S, Morisue T, Kobayashi T (1990) Biotech. Bioeng. 36:520
52. Itoh T, Honda H, Shimizu K, Kobayashi T (1991) Appl. Microbiol. Biotechnol.
53. Cohon JL (1978) Multiobjective programming and planning. Academic, New York
54. D'Ans G, Gottlieb D, Kokotovic PV (1972) Automatica 8:729
55. D'Ans G, Kokotovic PV, Gottlieb D (1971) IEEE Trans. Auto. Cont. AC-16:341
56. Muzychenko LA, Aascheva LA, Yakovleva GV (1974) Biotechnol. Bioeng. Symp. No. 4:629
57. Yamane T, Sada E, Takamatsu T (1979) Biotech. Bioeng. 21:111
58. Takamatsu T, Hashimoto I, Shioya S, Mizuhara K, Koike T, Ohno H (1975) Automatica 11:141
59. Kalman RE, (1960) Trans. ASME J Basic Eng. 82:35
60. Kalman RE, Bucy RS (1961) Trans. ASME J Basic Eng. 108:95
61. Svrcek WY, Elliot RF, Jajic JE (1974) Biotech. Bioeng. 16:827
62. Bellgart KH, Kuhlman W, Meyer HD, Schügerl K, Thoma M (1986) IEE Proc. 133:226
63. Nahlic J, Buriance Z (1988) Appl. Microbiol. Biotechnol. 28:128
64. Ghoul M, Pons MN, Engasser JM, Bordet (1985) Proc. of IFAC Modeling and control of biotechnological processes, Noordwijkerhout: 185
65. Stephanopoulos G, San KY (1984) Biotech. Bioeng. 26:1176
66. Shimizu H, Takamatsu T, Shioya S, Suga K (1989) Biotech. Bioeng. 33:354
67. Mou DG, Cooney CL (1983) Biotech. Bioeng. 25:225
68. Bastin G, Dochain D (1986) Automatica 22:705
69. Dochain D, Pauss A (1988) The Canadian J. of Chem. Eng. 66:626
70. Jazwinski AH (1970) Stochastic processes and filtering theory. Academic, New York
71. Ramirez WF (1987) Chem. Eng. Sci. 42:2749
72. Aborhey S, Williamson D (1978) Automatica 14:493
73. Holmberg A, Ranta J (1982) Automatica 18:181
74. Dochain D, Bastin G (1984) Automatica 20:621
75. Aiba S (1979) Biotechnol. Bioeng. Symp. No. 9:269
76. Cooney CL (1979) Biotechnol. Bioeng. Symp. No. 9:1

77. Yamashita S, Hoshi H, Inagaki T (1969) In: Fermentation Advances, Perlman D (ed) Acad. Press: 441
78. Grayson P (1969) Proc. Biochem. 3:43
79. Nyiri LK (1972) In: R. Squires (ed) Development in industrial microbiology. vol 13, Fort Collins, Washington DC
80. Rolf MJ, Hennigan PJ, Mohler RD, Weigand WA, Lim HC (1982) Biotech. Bioeng. 24:1191
81. Mohler RD, Hennigan PJ, Lim HC, Tsao GT, Weigand WA (1979) Biotechnol, Bioeng. Symp. No 9:257
82. Wang HY, Cooney CL, Wang DIC (1979) Biotech. Bioeng 21:975
83. Huang SY, Chu WB (1981) Biotech. Bioeng. 23:1491
84. Yano T, Kobayashi T, Shimizu S (1981) J. Ferment. Technol. 59:295
85. Nyeste L, Szigeti L, Veres A, Pungor Jr E, Kurucz I, Hollo J (1981) Biotech. Bioeng. 23:405
86. Verez A, Byeste L, Kurcz I, Kirchknopf L, Szigeti L, Hollo J (1981) Biotech. Bioeng. 23:391
87. Dairaku K, Izumoto E, Morikawa H, Shioya S, Takamatsu T (1983) J. Ferment. Technol. 61:189
88. Omstead DR (ed.) (1990) Computer control of fermentation processes. CRC Press, Boca Raton, Florida.
89. Leigh JR (ed) (1987) Modelling and control of fermentation processes. Peter Peregrinus, United Kingdom.
90. Johnson A (1987) Automatica 23:691
91. Cardello RJ, San KY (1988) Biotech. Bioeng. 32:519
92. Takamatsu T, Shioya S, Okada Y, Kanda M (1985) Biotech. Bioeng. 27:1675
93. Shimizu H, Shioya S, Suga K, Takamatsu T (1989) Appl. Microbiol. Biotechnol. 30:276
94. Shimizu K, Morikawa M, Mizutani S, Iijima S, Matsubara M, Kobayashi T (1988) J. Chem. Eng. Japan 21:113
95. Wu WT, Chen KC, Chiou HW (1985) Biotech. Bioeng. 27:756
96. Williams D, Yousefpour P, Wellington EMH (1986) Biotech. Bioeng. 28:631
97. Radjai MK, Hatch RT, Cadman TW (1984) Biotech. Bioeng. symp. No. 14:657
98. Dekkers RM, Voetter M (1985) Proc. of IFAC Modeling and Control of Biotechnological Processes, Noordwijkerhout, 103
99. Verbruggen HB, Eelderink GHB, Van den Broecke PM (1985) Proc, of IFAC Modeling and Control of Biotechnological Procresses, Noordwijkerhout: 121
100. Frueh K, Lorenz Th, Niehoff, Diekmann J, Hiddessen R, Schügerl K (1985) Proc. of IFAC Modeling and Control of Biotechnological Processes, Noordwijkerhout, p 75
101. Yust LJ, Howell JA (1988) Chem. Eng. Res. Des. 66:260
102. Montague GA, Morris AJ, Wright AR, Aynsley M, Ward A (1986) Can. J. of Chem. Eng. 64:567
103. Clark DW, Mohtadi C (1985) Self-tuning control of a difficult process: 7th Symposium on Identification and System Parameter Estimation, University of York, UK
104. Cutler CR, Ramaker BL (1979) Dynamic Matric Control-A Computer Control Algorithm: Preprint for the 86th AIChE National Meeting, Houston, TX
105. Goochee CF, Hatch RT, Cadman TW (1989) Biotech. Bioeng. 33:282
106. Hunt LR, Su R, Meyer G (1983) Design for Multi-input Nonlinear Systems. In: Brockett RW et al. (eds) Differential geometric control theory p 268
107. Hoo KA, Kantor JC (1986) Math. Biosci. 82:43
108. Ollis DF (1984) Competition between two species when only one has antibiotic resistance, Chemostat Analysis, Paper Presented at AIChE Meeting, San Francisco Nov.
109. Henson MA and Seborg DE (1992) Chem. Eng. Sci. 47:821
110. Yamagata Y. Shimizu K, Iijima S, Morisue T, Kobayashi T (1989) An efficient operation of Bioreactor Syatems by Repetitive Learning Control strategy, in preparation
111. Shimizu H, Sada E, Shioya S, Suga K (1989) Biotech. Bioeng. 34:794
112. Kishimoto M, Yoshida T, Taguchi H (1981) J. Ferment. Technol. 59:125
113. Bamberger W, Iserman R (1978) Automatica 14:223
114. Rolf MJ, Lim HC (1984) Chem. Eng. Commun. 29:229
115. Rolf MJ, Lim HC (1985) Biotech. Bioeng. 27:1236
116. Semones GB, Lim HC (1989) Biotech. Bioeng. 33:16
117. Shi Z, Shimizu K, Watanabe N, Kobayashi T (1989) Biotech. Bioeng. 33:999
118. Shi Z, Shimizu K, Iijima S, Morisue T, Kobayashi T (1990) Biotech. Bioeng. 36:520
119. Kalogerakis N, Linardos T, Behie LA, Svrcek WY (1986) Can. J. Chem. Eng. 64:581

120. Boyle TJ (1979) Biotech. Bioeng. Symp. No. 9:349
121. Endo I, Asama H, Nagamune T (1989) A database system and an expert system for realizing factor automation in the bioindustries. In: Fiechter et al. (eds) Bioproducts and bioprocesses. Springer, Berlin Heidelberg New York
122. Asama H, Nagamune T, Endo I, Hirata M, Hirata A (1991) Kagaku Kogaku Ronbunshu (Japanese) 17:579
123. Nakamura T, Kuratani T, Morita Y (1985) Proc. of IFAC Modeling and control of biotechnological processes, Noordwijkerhout: 231
124. Fu CS, Wang SQ, Wang JC (1988) Proc. of the 4th Int. Conf. on Comp. Appl. in Ferm. Tech., Cambridge, 411–420
125. Chen Q, Wang SQ, Wang JC (1988) Proc of the 4th Int. Conf. on Comp. Appl. in Ferm. Tech., Cambridge, 253–261
126. Konstantinov K, Yoshida T (1989) Biotech. Bioeng. 33:1145
127. Yamada Y, Haneda K, Murayama S, Shiomi S (1991) J. Chem. Eng. Japn, 24:94
128. Oishi K, Tominaga M, Kawato A, Abe Y, Imayasu S, Nanba A (1991) J. Ferm. Tech. 72:115
129. Matsuura K, Hirotsune M, Nakada F, Hamachi M (1991) Hakkoukougaku, 69:455
130. Kishimoto M, Kitta Y, Takeuchi S, Nakajima M, Yoshida T (1991) J. Ferm. Tech. 72:110
131. Shi Z, Shimizu K (1992) J. Ferm. Bioeng. 74:39
132. Konstantinov KB, Yoshida T (1992) Biotech. Bioeng. 39:479
133. Gehlen S, Bettenhausen KD (1991) Modelling of biotechnological processes interpolating associative memories. In: Hanus R, Kool P, Tzafestas (eds) Mathematical and intelligent models in system simulation. Sci. Pub. Co., p 755
134. Gehlen S, Tolle H, Kreuzig J, Friedl P (1992) Integration of expert systems and neural networks for the control of fermentation processes. In: IFAC Symposium on Modeling and control of Biotechnical Processes. 29 March–2 April, Keystone, Colorado.
135. Bailey JE (1991) Science 252:1668
136. Stephanopoulos G and Vallino JJ (1991) Science 252:1675

The Principle and Technology of Hydrogen Peroxide Based Biosensors

John H. T. Luong[1], A.-L. Nguyen[1], and George G. Guilbault[2]

[1] Biotechnology Research Institute, National Research Council Canada, Montreal, Quebec, Canada H4P 2R2

[2] Department of Chemistry, University of New Orleans, New Orleans, Louisiana, USA

Oxidases can be used with a hydrogen peroxide electrode to construct amperometric biosensors for determination of various analytes, including substrates, cofactors, prosthetic groups, inhibitors, activators, enzyme activities, as well as antigens/antibodies. This chapter reviews the principle and technology of hydrogen peroxide-based biosensors and the techniques for enzyme immobilization. Methods for alleviation of interference by electroactive species are cited with an emphasis on mediator technology. The present state of applications and development trends of this technology are also presented and discussed.

Advances in Biochemical Engineering/
Biotechnology, Vol. 50
Maniging Editor: A. Fiechter
© Springer-Verlag Berlin Heidelberg 1993

1 Introduction

A biosensor is an analytical device that brings together an immobilized biological sensing material and a transducer to produce a digital electronic signal that is proportional to the concentration of a target chemical substance (Fig. 1). Biosensors also include bioprobes, sensors used for in vivo monitoring.

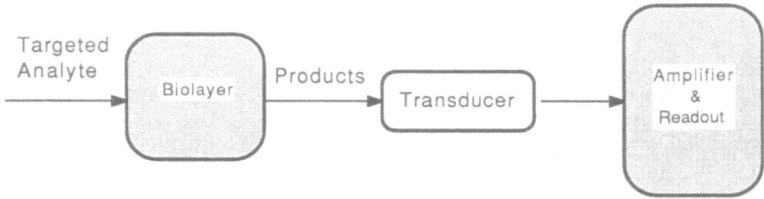

Fig. 1. Schematic diagram of a biosensor system in which an immobilized biosensing element is placed in intimate contact with a transducer

Biosensors are not exactly new Clarke and Lyons (1962) [1] described this concept using a soluble glucose oxidase which was retained on the surface of an amperometric oxygen electrode by a polymer membrane. As the enzyme reaction proceeded, the detection of oxygen consumption was followed and equated to the analyte concentration. In 1965, Kadish and Hall also described a similar system for measurement of glucose using soluble glucose oxidase with an oxygen electrode [2]. Updike and Hicks [3] took this idea a step further in 1967 by immobilizing glucose oxidase in a polyacrylamide gel and placing the resulting immobilized enzyme in intimate contact with an oxygen electrode. They christened this device "the enzyme electrode" and since then this terminology has gained wide popularity and acceptance. Following such papers, researchers have developed a vast array of biosensors based on a variety of different biodetecting elements and transducers (Table 1) [4, 5]. The specificity is the raison d'être for the widespread use of enzymes in biosensor construction. Together with oxidase systems, amperometric electrodes based on either oxygen or hydrogen peroxide measurements have been developed for the determination of about 80 different substances including substrates, co-factors, prosthetic groups, enzyme activities, antibodies/antigens, inhibitors, and activators [6]. Combining biology, physics, and electronics, the biosensor is now picking up steam as more and more applications come into view and onto the market. These applications are appearing in the realms of human and animal health care, environmental monitoring and pollution control, food processing and industrial bioprocessing, and defense, in which the primary use of biosensors is to detect toxins in biological and chemical warfare [7].

There are several key advantages to amperometric electrodes that make them very useful in biosensor development. First, amperometric electrodes are inexpensive and relatively simple in production and application. Second, they can operate

Table 1. Biosensor components

Biodetection elements	Transducers
Organisms	Potentiometric
Tissues	Amperometric
Whole cells	Conductometric
Organelles	Impedimetric
Membranes	Fibre optic
Enzymes	Calorimetric
Receptors	Acoustic/Piezoelectric
Antibodies	Molecular electronic
Nucleic acids	
Organic molecules	

over a wide concentration range, encompassing the range of interest for many practical applications. Under appropriate conditions concentrations of 10^{-6} to 10^{-9} M can be detected and a dynamic range of three to four orders of magnitude can be readily achieved. Third, a colored sample does not interfere in an electrochemical analysis, unlike spectrophotometric assays. Amperometric biosensors are usually based on oxidoreductase enzymes which commonly use oxygen, nicotinamide adenine dinucleotide NAD^+ or nicotinamide adenine dinucleotide phosphate $NADP^+$ as the electron acceptor, which recycles the enzyme after the substrate reaction. Enzymes using oxygen as the electron acceptor are referred to as oxidases while those using $NAD^+/NADH^+$ are called reductases or dehydrogenases. Oxidases are, by far, the most commonly used since they require O_2 instead of $NAD^+/NADH$ or $NADP^+/NADPH$ which are not always completely reversible. In addition when the reduced forms are reoxidized, some products can form and foul the electrode. After repeated use, the oxidized co-factor decreases in concentration and becomes a limiting factor. The electrochemistry of O_2/H_2O_2 is less problematical than that of $NAD(P)^+/NAD(P)H$ and oxygen is already available in most measured samples. This chapter will review the principle and technology of hydrogen peroxide-based biosensors using oxidase enzymes and the present state of applications. The potentials and limitations will be presented and the future of this technology discussed.

2 Principle of Hydrogen Peroxide Measurement

2.1 Enzymatic Oxidation

The overall reaction catalyzed by an oxidase can be described as follows:

$$XH_2 + Y \xrightarrow{\text{Oxidase}} X + H_2Y \tag{1}$$

If the immediate acceptor, Y, is oxygen, H_2O_2 will be formed in this reaction. Consequently, the amount of an analyte present in the solution could be determined by following either the rate of oxygen consumption or the rate at which hydrogen

Table 2. Oxidases useful in biosensor construction

Enzymes	EC	Sources	Substrate(s)	Reaction Products
Acetylindoxyl ox.	1.7.3.2	Plants	N-Acetylindoxyl	N-Acetylisation, H_2O_2
Alcohol ox.	1.1.3.13	Yeast, fungi	Primary alcohol	Aldehyde, H_2O_2
Aldehyde ox.	1.2.3.1	Animal liver	Aldehyde	Acid, H_2O_2
Amine ox.	1.4.3.4	Animal tissues, plants	Amine	Aldehyde, NH_3, H_2O_2
D-Amino acid ox.	1.4.3.3.	Liver, kidney, molds, bacteria	D-Amino acids	2-Oxoacid, NH_3, H_2O_2
L-Amino acid ox.	1.4.3.2	Liver, kidney, snake venom, molds, bacteria	L-Amino acids	2-Oxoacid, NH_3, H_2O_2
2-Aminophenol ox.	1.10.3.4	*Pycnoporus coccineus* *Bauhenia monandra*	2-Amino phenol	2-Quinoimine, H_2O_2
Aryl-alcohol ox.	1.1.3.7	White rot fungi	Aromatic prim. alcohol	Aromatic aldehyde, H_2O_2
D-Aspartate ox.	1.4.3.1	Kidney	D-Aspartate	Oxalocetate, NH_3, H_2O_2
Cholesterol ox.	1.1.3.6	Bacteria	Cholesterol	4-Cholesten-3-one, H_2O_2
Choline ox.	1.1.3.17	*Arthrobacter globiformis*	Choline	Betaine aldehyde, H_2O_2
Cyclohexylamine ox.	1.4.3.12	*Pseudomonas* sp.	Cyclohexylamine and some other cyclic amines	Cyclohexanone, NH_3, H_2O_2
Diamine ox.	1.4.3.6	Animal plasma, Plea seedlings	Adiamine	An aminoaldehyde, H_2O_2, NH_3
Dihydro-orotate ox.	1.3.3.1	*Zymobacterium oroticum* C	L-5,6-dihydro-orotate	Orotate, H_2O_2
Ecdysone ox.	1.1.3.16	*Calliphora erythrocephala*	Ecdysone,	3-Dehydroecdysone, H_2O_2
Ethanolamine ox.	1.4.3.8	*Arthrobacter*	Ethanolamine	Glycoaldehyde, NH_3, H_2O_2
Galactose ox.	1.1.3.9	Molds	D-Galactose	D-Galactohexodialdose, H_2O_2
Glucose ox.	1.1.3.4	Honey bee, molds	β-D-Glucose	Gluconic acid, H_2O_2
D-Glutamate ox.	1.4.3.7	Invertebrate tissue	D-Glutamate	2-Oxoglutamate, NH_3, H_2O_2
L-Glutamate ox.	1.4.3.11	Bacteria (*Azotabacter vinelandii*)	L-Glutamate	2-Oxoglutamate, NH_3, H_2O_2
Glutathione sulfhydryl ox.	n.a.	*Penicillum* sp.	GSH	GSSG, H_2O_2
Glycollate ox.	1.1.3.1	Animal tissue, plants	Glycollate	Glyoxylate, H_2O_2
Glyoxylate ox.	1.2.3.5	Bacteria	Glyoxylate	Oxalate, H_2O_2
L-Gulonolactone ox.	1.1.3.8	Animal tissues	L-Gulono-γ-lactone	L-Xylo-hexulolactone H_2O_2
Hexose ox.	1.1.3.5	Red alga	β-D-Glucose	D-Glucono-δ-Lactone, H_2O_2
L-2-Hydroxy acid ox.	1.1.3.15	Liver	L-2-Hydroxy acid	2-Oxoacid, H_2O_2

Enzyme	EC No.	Source	Substrate	Products
3-Hydroxyanthranilate ox.	1.10.3.5	Liver	3-Hydroxyanthranilate	1,2-Benzoquinoneimine-3-carboxylate, H_2O_2
6-Hydroxy-D-nicotine ox.	1.5.3.6	Arthrobacter oxidans	6-Hydroxy-D-nicotine	[6-Hydroxypyridyl(3)]-(3-N-methylamino propyl)-ketone, H_2O_2
6-Hydroxy-L-nicotine ox.	1.5.3.5	Arthrobacter oxidans	6-Hydroxy-L-nicotine	Ibid.
Lactate ox.	1.1.3.2	Pediococcus sp., Mycobacterium phlei	L-Lactate	Pyruvate, H_2O_2
Lathosterol ox.	1.3.3.2	Liver	5α-Cholest-7-en-3β-ol	Cholesta-5,7-dien-3β-ol, H_2O_2
L-Lysine ox.	1.4.3.14	Trichoderma viride	L-Lysine	α-Keto-ε-aminocaproate + NH_3 + H_2O_2
Malate ox.	1.1.3.3	Bacteria	L-Malate	Oxaloacetate, H_2O_2
N-Methylamino acid ox.	1.5.3.2	Kidney	N-Methyl-L-amino acid	L-Amino acid, HCHO, H_2O_2
N^6-Methyl-lysine ox.	1.5.3.4	Kidney	N^6-Methyl-L-lysine	L-Lysine, HCHO, H_2O_2
Monoamine ox.	1.4.3.4	Beef plasma, Human placenta	Monoamine	Aldehyde, H_2O_2
Nitroethane ox.	1.7.3.1	Molds	Nitroethane	Acetaldehyde, nitrile, H_2O_2
Oxalate oxidase	1.2.3.4	Mosses	Oxalate	CO_2, H_2O_2
Putrescine ox.	1.4.3.10	Micrococcus rubens	Putrescine	4-Aminobutyraldehyde, NH_3, H_2O_2
Pyranose ox.	1.1.3.10	Fungi	D-Glucose	D-Glucosone, H_2O_2
Pyridoxamine phosphate ox.	1.4.3.5	Animal tissue, yeast, bacteria	Pyridoxaminephosphate	Pyridoxalphosphate, NH_3, H_2O_2
Pyridoxine ox.	1.1.3.12	Pseudomonas	Pyridoxine	Pyridoxal, H_2O_2
Pyruvate ox.	1.2.3.3	Lactobacillus	Pyruvate, orthophosphate	Acetyl phosphate, CO_2, H_2O_2
Pyruvate ox. (CoA-acetylating)	1.2.3.6	Bacteria	Pyruvate, CoA	Acetyl-CoA, CO_2, H_2O_2
Sarcosine ox.	1.5.3.1	Liver	Sarcosine	Glycine, HCHO, H_2O_2
L-Sorbose ox.	1.1.3.11	Trametes sanguinea	L-Sorbose	5-Keto-D-fructose, H_2O_2
Spermine ox.	1.5.3.3	Neisseria perflava, Sarratia marcescens	Spermine, spermidine	H_2O_2
Sulfite ox.	1.8.3.1	Animal tissues, plant, bacteria	Sulfite	Sulfate, H_2O_2
Tyramine ox.	1.4.3.9	Sarcina lutea	Tyramine	4-Hydroxyphenyl acetaldehyde, NH_3, H_2O_2
Urate ox.	1.7.3.3	Animal tissues, yeast	Urate	Allantoin, H_2O_2
Xanthine ox.	1.2.3.2	Animal tissues, milk, bacteria	Xanthine/hypoxanthine. Also oxidizes other purines, pterins and aldehyde.	Uric acid, H_2O_2

peroxide is produced. A classical example of this type of reaction is the oxidation of glucose to gluconic acid by glucose oxidase (GO).

$$\text{glucose} + O_2 \xrightarrow{\text{GO}} \text{gluconic acid} + H_2O_2 . \qquad (2)$$

Apparently, an enzyme-based biosensor can be constructed for any analyte provided a suitable oxidase is available. To date, several oxidases have been identified and/or isolated (Table 2) [8]. The preparation of a practical enzyme based biosensor, however, requires that the enzyme must fulfill the following criteria: a) high selectivity, b) good stability, and c) high specific activity. The selectivity of an oxidase is a sine qua non since it will affect the selectivity of the biosensor. For instance, glucose oxidase obtained from *Penicillium* is highly specific for β-D-glucose [9]. Any alteration whatsoever in this molecule enormously reduces the rate of oxidation. The high selectivity of this enzyme makes it one of the most useful reagents for the detection and estimation of glucose in biological fluids. Xanthine oxidase, however, has low selectivity and attacks a number of aldehydes, purines, pyrimidines, pteridines, azopurines, and other heterocyclic compounds [10]. Similarly, alcohol oxidase from *Basidiomycetes* sp. oxidizes a number of alcohols including methanol, ethanol, n-propanol, n-butanol, and allyl alcohol [11, 12]. This enzyme also responds to L-lactic acid, acetic acid, formic acid, butyric acid, formaldehyde and aldehyde [13]. Stability of the enzyme is another matter of concern since it dictates the operating life of the biosensor. There is a general belief that immobilization leads to an improvement in stability of the immobilized enzymes towards physical and chemical stress. The immobilized enzyme appears to be less susceptible to the normal activators and inhibitors that affect the soluble enzyme. For instance, alcohol oxidase is stable for only one week or less in soluble forms but becomes very stable (> 4 months) after being immobilized [13]. However, under some circumstances, the immobilized preparation may show a decrease in stability. Chibata [14] reports that of 50 enzymes immobilized, 60% showed an increase in stability, 16% showed a decrease, and 21% were unaffected. Consequently, prediction of immobilized enzyme properties is somewhat uncertain.

An enzyme with a high specific activity is always desirable for the development of a fast response biosensor. Otherwise, a considerably high amount of enzyme loading is necessary to achieve satisfactory performance. For example, commercial Sigma glucose oxidase obtained from *Aspergillus niger* may contain up to 200 IU mg^{-1} protein. Alcohol oxidase (Sigma) prepared from *Pichia pastoris*, however contains only 10–40 IU mg^{-1} protein. To be useful in an electrode, the activity should be at least 10 IU mg^{-1} protein [15].

2.2 Amperometric Detection of Hydrogen Peroxide

Electrochemical measurement is categorized by its modus operandi: amperometric, potentiometric, and conductometric. Conductometric measurement is not widely used since it is non-selective and exhibits poor signal-to-noise ratios. However, conductometric biosensors may gain more attention in the future because of their simple production and application [16]. Potentiometric sensors (ion-selective electrode or gas-sensing electrode) are selective and exhibit wide response ranges but have the disadvantage of sluggish responses. Amperometry is preferred to potentiometry since the requirement of an accurate reference electrode is obviated.

In amperometric electrodes, a constant potential is applied between the sensing electrode and a reference electrode. The current generated by the redox reaction of the analyte at the sensing electrode is directly proportional to the analyte concentration at the electrode surface (Fig. 2). Since the analyte is consumed during the process, the measurements are critically dependent on the rate of analyte transport to the electrode surface. For H_2O_2 measurement, an electrode usually made of platinum is poised at $+500$ to $+800$ mV (i.e., the anode) relative to the Ag/AgCl reference. Under this condition, the electrode is completely insensitive to oxygen (potential applied for oxygen detection is ca. -0.7 V) but responds to hydrogen peroxide and is electrochemically oxidized as follows:

$$H_2O_2 \rightarrow 2\,H^+ + O_2 + 2\,e^- \cdot \tag{3}$$

Other electrode constructions of Au and various forms of carbon have been used as an alternative. Various preparation steps involving polishing, heat treatment, and cycling of the electrode between several different potentials have improved both reproducibility and response [17].

The reference can be made from silver, or more commonly, silver coated with silver chloride (Ag/AgCl). This is a pseudo-reference electrode since there is no internal filling solution or electrolytic junction and the electrode contacts the sample solution directly. For Ag/AgCl the potential of the reference electrode

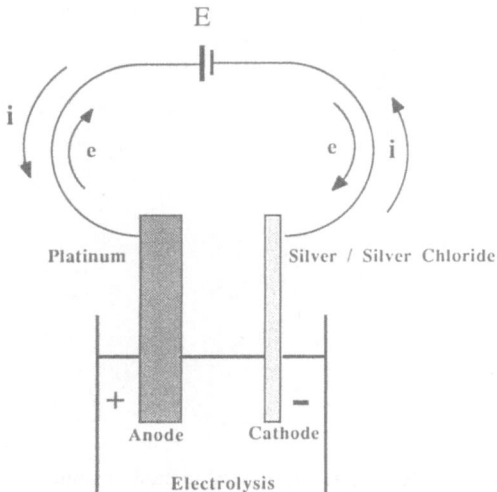

Platinum   Silver / Silver Chloride

Anode   Cathode

Electrolysis

Fig. 2. Principle of electrolysis: At the platinum (A), electrons are removed from the substrate ($H_2O_2 \rightarrow 2\,H^+ + 2\,e^- + O_2$) and oxidation occurs. At the silver/silver chloride cathode (C), electrons are added and reduction occurs ($4\,H^+ + O_2 + 4\,e^- \rightarrow 2\,H_2O$)

depends on the chloride ion concentration in the sample solution. As long as the chloride ion level is constant, the response electrode potential remains unchanged and the performance of this pseudo-reference electrode is similar to that of a true reference electrode. Any change in chloride ion, will lead to a change of reference potential, which in turn changes the working electrode potential and results in a different current response. The construction of the pseudo-reference electrode is greatly simplified but the measurement is dependent on the chloride ion concentration of the sample solution.

At the silver cathode, oxygen is reduced to water by the following reaction:

$$4\,H^+ + O_2 \rightarrow 2\,H_2O - 4\,e^-\,. \tag{4}$$

The overall electrochemical oxidation of hydrogen peroxide can thus be expressed as:

$$H_2O_2 + 2\,e^- + 2\,H^+ \rightarrow 2\,H_2O\,. \tag{5}$$

The advance of amperometric techniques from the early two-electrode to three-electrode systems has increased sensitivity and accuracy [18]. In classical instrumentation with the two-electrode system, the potential is applied across the working electrode. Current flow through a cell of high resistance causes an appreciable voltage drop. Thus, the potential at the working electrode differs from that at the other end of the cell. Modern amperometric instrumentation employs a third electrode, a reference electrode of constant potential is positioned as close as possible to the working electrode to sense the potential at that point (Fig. 3).

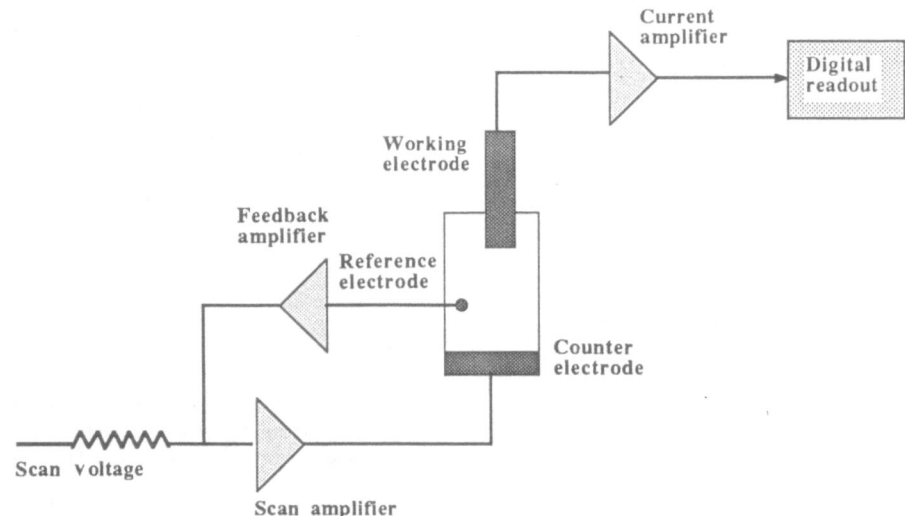

Fig. 3. Schematic diagram of three-electrode potentiostat system used in modern amperometric instrumentation

Based on this measurement, a corrective voltage will be applied to maintain a constant potential to within 1 mV for long periods. The reference electrode is loaded only to the extent of picoamperes (10^{-12} A) or less [18]. Consequently, the potential of the working electrode of the three-electrode system can be accurately controlled regardless of solution conditions, whereas, the potential of the working electrode in a two-electrode system is dependent on solution conditions.

Hydrogen peroxide sensors have a good sensitivity and linear range. However, they lack the selectivity because they respond to any species that is electroactive at the working potential. Substances such as uric acid, ascorbic acid, glutathione, cysteine, acetaminophen, bilirubin, tyrosine, etc. are known as potent interfering substances. The problem of interference will be discussed later, together with the solution to overcome such a drawback.

3 Enzyme Immobilization

Most enzymes are very expensive by comparison with other reagents, and routine analysis requires large and costly amounts of such materials. Immobilized enzymes can be reused several times, even up to 10000 times in some cases, thus representing a tremendous cost saving. The three following methods have been frequently employed to immobilize enzymes used in hydrogen peroxide electrodes:
1. Covalent binding to small water-insoluble particles
2. Covalent binding to water-insoluble membranes
3. Enzyme retention by an electropolymerized film.

3.1 Covalent Binding to Small Water-Insoluble Particles

Enzymes can be covalently immobilized on water-insoluble particles such as porous glass, polyacrylamide, polyacrylic acid derivatives, polyaspartic acid, polyglutamic acid, polystyrene, nylon, cellulose, Sephadex, ethylene-maleic anhydride copolymer, Agarose, Sepharose, carboxymethyl cellulose, etc. [19]. Covalent binding has the great advantage in that, leakage of the biological components is unlikely to occur during use of the biosensor. Binding is accomplished through functional groups in the enzyme which are not essential for its catalytic activity. Reaction preferably takes place at low ionic strength, low temperature, and within the optimal pH range of the soluble enzyme. Often coupling is carried out in the presence of the enzyme's substrate or the enzyme's inhibitor to protect the active site. The amino acid residues suitable for covalent binding are: (1) α- and δ-amino groups, (2) the phenol ring of tyrosine, (3) β- and γ-carboxyl group, (4) the sulfhydryl (thiol) group of cysteine, (5) the hydroxyl group of serine, and (6) the imidazole group of histidine. Of these, the most widely used are the first three (Table 3). The water-insoluble carriers are usually activated by transformation into various derivatives. There are a large number of methods of covalently coupling enzymes

Table 3. Some covalent immobilization procedures useful in biosensor construction

Carrier function	Reagent	Active intermediate	Product
Via Amino Group			
—OH / —OH	BrCN		
—OH	 X = Cl, NH₂, OCH₂COOH		
—CH₂—OH			—CH₂—NH—Ⓔ
—CH₂—NH₂	CHO(CH₂)₃CHO	—CH₂N=CH—(CH₂)₃—CHO	

Functional Group	Reagent	Intermediate	Product
$-CN$	C_2H_5OH/HCl	$-\underset{\parallel}{C}-O-C_2H_5$ ($=NH$)	$-\underset{\parallel}{C}-NH-Ⓔ$ (NH)
$-COOH$	$\underset{N-R}{\overset{}{N=C=N-R'}}$	$-CO_2-\underset{=N-R'}{\overset{NH-R}{C}}$	$-CO-NH-Ⓔ$
$-NH-CH-CH_2$ / $O=C-S$	None	N/A	$-NH-CH-\underset{\parallel}{C}-NH-Ⓔ$ (CH_2SH / $C=O$)
Via Hydroxyl Group			
$-CH_2OH$	$TiCl_4$	$-CH_2OTiCl_3$	$-CH_2OTiCl_2-O-Ⓔ$
Via Phenol Ring			
$-NH_2$	$NaNO_2/HCl$	$-\overset{Cl^-}{\underset{}{N\overset{+}{\equiv}N}}$	(phenol ring: OH, $-N=N-$, $-CH_2-Ⓔ$)

Table 3 (Continued)

Via Carboxyl Group

$\|-Y$ $Y = NH_2$ or SH	$N=C=N$ with $-R$, $-R'$	$Z=C$ with $NH-R$, $N-R'$ $Z = NH$ or S	$\|-Z-C(=O)-\text{\textcircled{E}}$

Via Thiol Group

$\|-SH$	$\big\langle\!\!\big\rangle\!-S-S-\!\big\langle\!\!\big\rangle$	$\|-S-S-\!\big\langle\!\!\big\rangle$	$\|-S-S-\text{\textcircled{E}}$
$\|-NH_2$	$NaNO_2/HCl$	$\|-N^+\!\equiv N \;\; Cl^-$	$\|-N=N-S-\text{\textcircled{E}}$
$\|-OH$	$\underset{\displaystyle CH_2-CH-CH_2-Cl}{\overset{O}{\diagup\diagdown}}$	$\underset{\displaystyle O-CH_2-CH-CH_2}{\overset{O}{\diagup\diagdown}}$	$\|-OH-CH_2-\underset{OH}{CH}-CH_2-S-\text{\textcircled{E}}$
$\|-CH_2-OH$	$R-SO_2Cl$ $R = CH_2CF_3,\ C_6H_4CH_3$	$\|-CH_2-O-SO_2-R$	$\|-CH_2-S-\text{\textcircled{E}}$

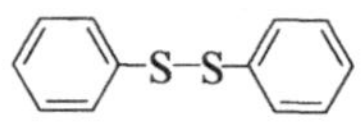

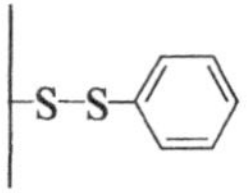

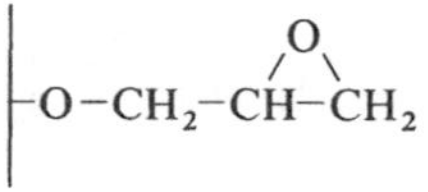

to water-insoluble carriers. Several review articles afford a good starting point for an appreciation of this field [20, 21, 22].

After immobilization, several changes may occur in the enzyme's apparent behavior. The optimal pH may shift, depending on the nature of the carrier. Often, the apparent activity of an enzyme is reduced after immobilization. Such behavior is usually related to the charge on the substrate and/or carrier, diffusion effects, and in some cases, tertiary changes in enzyme configuration.

The rate of inactivation and denaturation of an immobilized enzyme is less than that of the free enzyme. However, enzymes with excellent thermal stability do not necessarily possess excellent operational stability, since the latter depends on several other factors such as carrier durability, types of inhibitors and their concentration [19].

Immobilized enzymes with high apparent activity can be held between two membranes, one of which is placed in contact with a hydrogen peroxide sensitive electrode. The outer membrane can be made of polycarbonate material with a pore size large enough to pass analytes, oxygen, and other molecules but too small to allow macromolecules to pass through [23]. The inner membrane can be of cellulose acetate (cut off 100 Da) which only allows hydrogen peroxide or other small molecules to pass through (Fig. 4). A number of such proprietary combination membranes have been developed and commercialized. The Yellow Springs Instrument glucose and lactate membranes, are prepared in this way using a cellulose acetate and polycarbonate combination and the enzymes are immobilized on glutaraldehyde activated resin particles. This approach, however, is not applicable for immobilized enzymes with low apparent activity in view of sensitivity, response time and reusability. In this case, immobilized enzymes can be packed to form an enzyme column and are placed in close proximity to the hydrogen peroxide electrode. Practically, the enzyme column should allow at least many hundreds of separate analyses to be made before enzyme replacement is necessary. The commercialization of the Owens-Illinois glucose analyzer was developed based on this concept using glucose oxidase immobilized to a porous alumina support [24].

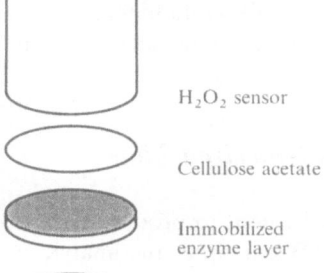

H_2O_2 sensor

Cellulose acetate

Immobilized enzyme layer

Polycarbonate membrane

Fig. 4. Immobilized enzyme layer is held between two membranes. The outer membrane with a large pore size allows the passage of small analytes and oxygen but not macromolecules. The inner membrane only allows hydrogen peroxide or other small molecules (MW < 40) to pass through

3.2 Covalent Binding to Water-Insoluble Membranes

Enzymes with high specific activity may also be covalently immobilized on a derivatized Teflon or nylon membrane and placed on the tip of the Clark-type hydrogen peroxide electrode. This method offers the advantage of simple preparation and linearity over a greater range of concentration. Several proprietary hydrophilic microporous membranes have been developed and a few of them are on the market. Immunodyne™ (Pall BioSupport, NY) is a specially modified, optically pure white nylon 66 microporous membrane. It features a chemically preactivated surface offering a high density of covalent binding sites that permanently immobilize enzymes/proteins. The immobilization is very simple and rapid as it only requires direct contact between the enzyme and the membrane [25]. This is a definite advantage as the procedures reported in the literature for the preparation of bioactive membranes are very tedious and time-consuming. Similarly, Immobilon™ affinity membrane (Millipore) is a chemically activated membrane to which a variety of enzymes/proteins can be irreversibly immobilized. The external and internal surface is chemically derivatized to offer a high capacity for covalent immobilization at pH 4–10.

Enzymes have also been immobilized onto collagen membranes by a number of different techniques. An improved, mild method for covalent coupling of enzymes including glucose oxidase was developed and patented by Coulet et al. [26] (Table 4). The enzymatic membrane has good resistance to deactivation by heat and denaturing reagents. This technique is incorporated into electrodes for glucose that are sold commercially by Tacussel (Lyon, France). The enzymic membrane is capable of at least 1000 assays in 3 months.

For enhancement of the enzyme loading, the enzyme can be cross-linked with a protein such as bovine serum albumin to form a thin protein layer on the membrane surface. Numerous bifunctional agents are available for this purpose including glutaraldehyde, hexamethylene disocyanate, 1,5-difluoro-2,4-dinitrobenzene, toluene-2-isocyanate, 4-isothiocyanate, bisdiazobenzidine-2,2'-disulfonic acid, and N-ethyl-5-phenylisoxazolium 3'-sulfonate (Woodward reagent K). Glutaraldehyde, the most common reagent, often reacts with the lysine amino groups of an enzyme. Such reactions must be optimized with respect to enzyme concentration, protein concentration, and glutaraldehyde level. The enzymic membrane prepared by cross-linked nucleoside phosphorylase (NP) with bovine serum albumin via glutaraldehyde was reported to have a 4–5 fold higher activity compared to that obtained when NP was directly immobilized covalently on the same membrane [27].

3.3 Enzyme Immobilization by an Electropolymerized Film

Immobilization on particles or membranes can only be used to produce enzyme electrodes having diameters in the millimeter range. When either the analyte or the sensing component is available in limited quantity, mini-electrodes are prerequisites for practical routine applications. Moreover, only micro-electrodes,

Table 4. Procedure for activation of collagen membranes

* Activation

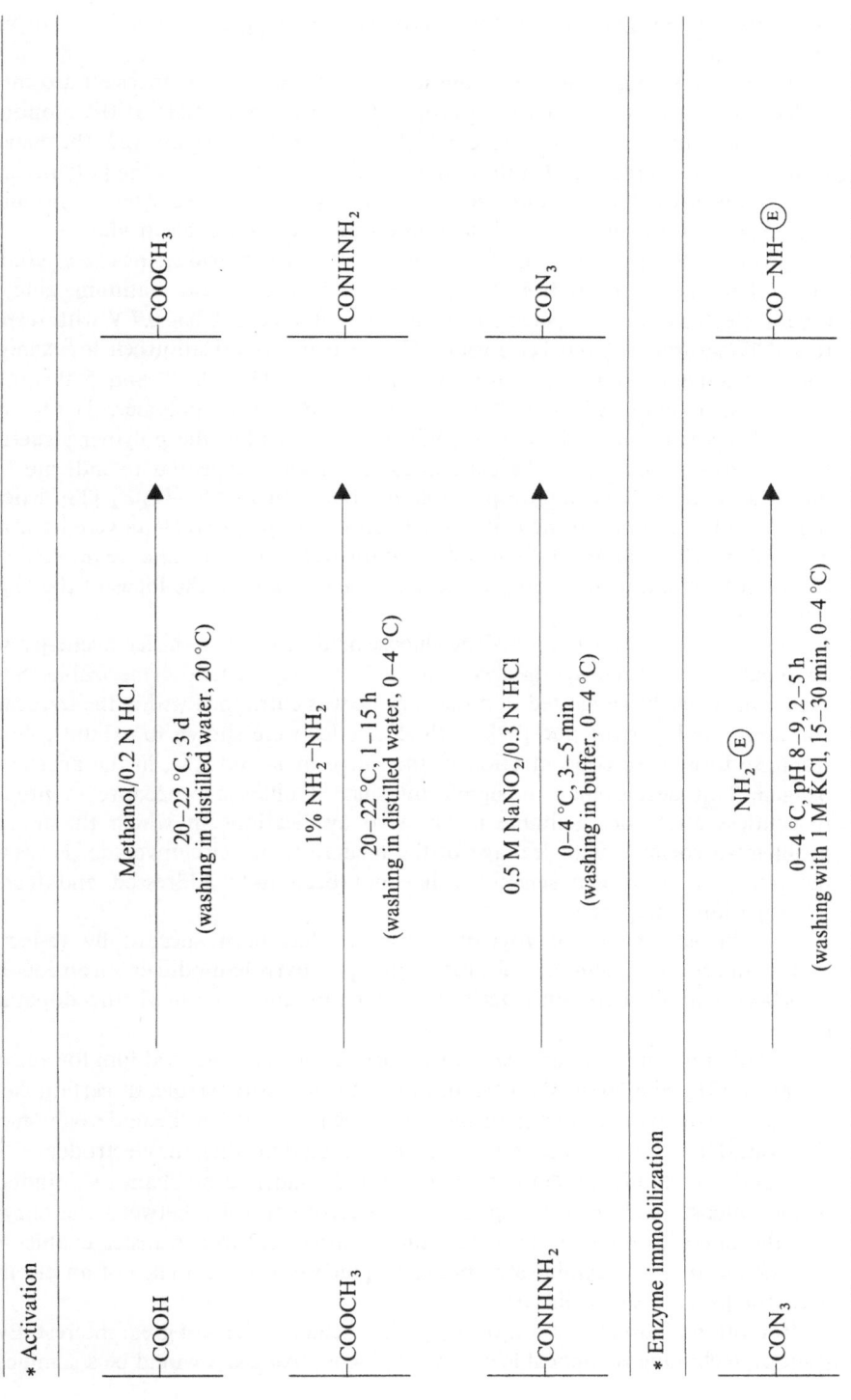

$-COOH \xrightarrow[\text{(washing in distilled water, 20 °C)}]{\substack{\text{Methanol/0.2 N HCl} \\ 20-22\,°C,\ 3\,d}} -COOCH_3$

$-COOCH_3 \xrightarrow[\text{(washing in distilled water, 0-4 °C)}]{\substack{1\%\ NH_2-NH_2 \\ 20-22\,°C,\ 1-15\,h}} -CONHNH_2$

$-CONHNH_2 \xrightarrow[\text{(washing in buffer, 0-4 °C)}]{\substack{0.5\ M\ NaNO_2/0.3\ N\ HCl \\ 0-4\,°C,\ 3-5\,min}} -CON_3$

* Enzyme immobilization

$-CON_3 \xrightarrow[\text{(washing with 1 M KCl, 15-30 min, 0-4 °C)}]{\substack{NH_2-\textcircled{E} \\ 0-4\,°C,\ pH\ 8-9,\ 2-5\,h}} -CO-NH-\textcircled{E}$

i.e., those having diameters in the micrometer range, are acceptable for in vivo applications.

Many monomers, in aqueous solutions, are known to polymerize onto metal surfaces that are maintained at appropriate positive potentials. If the monomer solution also contained an enzyme, which is often negatively charged, the enzyme would migrate to the metal surface and therefore, be trapped by the polymerizing film. The electropolymerization process effectively serves as an enzyme immobilization procedure to enable the fabrication of mini- and microelectrodes.

Pyrrole is the most often used monomer, and glucose oxidase has been used in most of the reported studies [28]. For electrodeposition, on platinum, gold, or carbon electrodes, the targeted electrode is maintained at 0.65–0.7 V with respect to a SCE (saturated calomel electrode). The electropolymerization cell, for example, contains aqueous solution of 0.35–0.4 M pyrrole, 0.1 M KCl, and 500 U ml^{-1} of glucose oxidase. Since polypyrrole is a conducting polymer, H_2O_2 may be oxidized at the underlying electrode surface or within the polymer structure. Under certain conditions, the experimental evidence appeared to indicate that only the metal surface was responsible for the reduction of H_2O_2 [28]. Microparticles of platinum, formed by immersion of the polypyrrole covered enzyme electrode in a hexachloroplatinate (V) solution, were found to enhance the response of the electrode, thus providing more evidence concerning the locus of the H_2O_2 reduction.

Since ferrocene is a mediator of the glucose oxidase activity, the ferrocene-pyrrole conjugates have been synthesized and were found electropolymerizable. Some preliminary results suggested that the enzyme was entrapped within the ferrocene-containing polypyrrole film [29]. Such electrodes were shown to exhibit a linear response to glucose concentration in the range of 1–100 mM, in the absence of oxygen. This development is significant since it offers a procedure to prepare reagentless electrodes, suitable for in vivo applications in which the toxicity problems associated with leakage of the mediators are circumvented. However, the question of oxygen sensitivity has not been fully addressed and further investigation is required.

The glucose oxidase-polypyrrole biosensor has been successfully tested in a flow-injection system. In addition, the polypyrrole-modified electrode has also been exploited to construct sensors for immunoglobulin G and dopamine [30].

N-methyl pyrrole can also be used to form electropolymerized film for enzyme retention [31]. Similar to the case of unsubstituted polypyrrole, in certain cases, evidence exists to prove that hydrogen peroxide is reduced at the underlying metal electrode. However, a separate study demonstrated that when the electrodeposition was conducted at higher temperature (50 °C), a conformation change was induced in the glucose oxidase resulting in direct electron transfer between the enzyme and the underlying electrode metal. Such a direct electron transfer enables the enzyme electrode to exhibit a response to glucose in the absence of an electron acceptor (oxygen or mediator).

Promotion of direct electron transport has been a subject of great interest. Even if such a technique is applicable only with glucose oxidase it would be a significant development [32]. Previously it has been demonstrated that direct electron transfer occurred between the glucose oxidase *co*-factor (the FAD group) and the (amino-phenyl) pyronic acid-modified glassy carbon electrode. Recently it was shown that a complex of osmium bis-pyridyl and polyvinylpyridine could be cross-linked by triethylenetetramine in the presence of glucose oxidase [33]. A gold electrode bearing such a GO containing cross-linked polymer responded linearly to glucose concentrations with a response time of less than 10 seconds.

Compared with the above-mentioned procedures, the electropolymerization of N-methyl pyrrole appears to be attractively simple, raising the possibility for the economical fabrication of microelectrodes. Unfortunately, the higher temperature electrodeposition was reported to adversely affect the selectivity of the enzyme [31], and further studies will be required to find an optimum protocol.

Tyramine, i.e. *para*(2-aminoethyl)-phenol, is another monomer that has been electro-polymerized onto metal electrodes [34]. After a blocking procedure, the electropolymerized film was not susceptible to non-specific adsorption of proteins, yet still possessed functional groups to which proteins could be chemically attached. An antibody of IgG was demonstrated to link to such a film, by action of a carbodiimide, resulting in an immunosensor capable of detecting IgG in the range of $10 \, pg \, ml^{-1}$ to $1 \, mg \, ml^{-1}$. Properties of the polytyramine film appeared to depend on the electrolysis conditions. Acidic solutions led to conductive films, while the film obtained from alkaline solutions acted as an insulator.

An earlier study reported on the modification of glucose oxidase by incorporating electron relays such as ferrocene derivatives, or azo compounds, into the glucose oxidase molecules [35]. Apparently the built-in relay centers promoted the direct electrical communication between the enzyme and the underlying electrode metal. The electropolymerized films likely can be used to entrap or to serve as binding supports for such modified enzymes, for the fabrication of microbio-sensors. Again, the problem of oxygen sensitivity must be investigated and better understood.

When the enzymes are immobilized on particles and held in place by an outer membrane, or immobilized directly on a membrane, the membrane could be appropriately chosen to screen out the interfering compounds. The microelectrodes, of course, also have to face the interference problem. Fortunately, many substituted phenylene oxides can be electropolymerized on different metals, including platinum and gold. By using the appropriate monomer(s), electropolymerized films could be formed with controlled porosity. Various compositions have been evaluated and the resorcinol-paradiaminobenzene film was found to afford the best protection and stability. The biosensor bearing glucose oxidase and protected by such a film was not sensitive to ascorbic acid, acetaminophene, or uric acid, and it could be used to monitor glucose levels in serum samples [36]. The same technique has also been employed to produce xanthine oxidase-bearing electrodes capable of quantitative determination of hypoxanthine/xanthine levels in fish extract [37].

4 Biosensor Methodology

Biosensor methodology can be divided into three types of measurements: batch, flow injection, and continuous.

4.1 Batch Measurement

In the commonest type of batch measurement, an enzyme is immobilized on a water-insoluble membrane which is attached to the sensing area of the hydrogen peroxide electrode. The electrode is mounted in a sample measurement chamber which is as small as 400 µl or less. The chamber contents are stirred by an air-driven silicone diaphragm. The detection chamber, of course, is provided with an inlet and an outlet for buffer and an injection port (Fig. 5). The measurements are made at steady-state response or the rate method where the slope of the response is measured after a certain time. Compared to the rate method or its variation, the steady-state measurement is slower, but following the complete response curve may indicate biosensor problems which would otherwise remain unnoticed.

4.2 Flow Injection

For enzymes with low specific activities, the resulting enzymic membrane may still exhibit very low activity. In this case, enzyme is preferably immobilized on water-insoluble particles to form an enzyme reactor column. The sample is injected

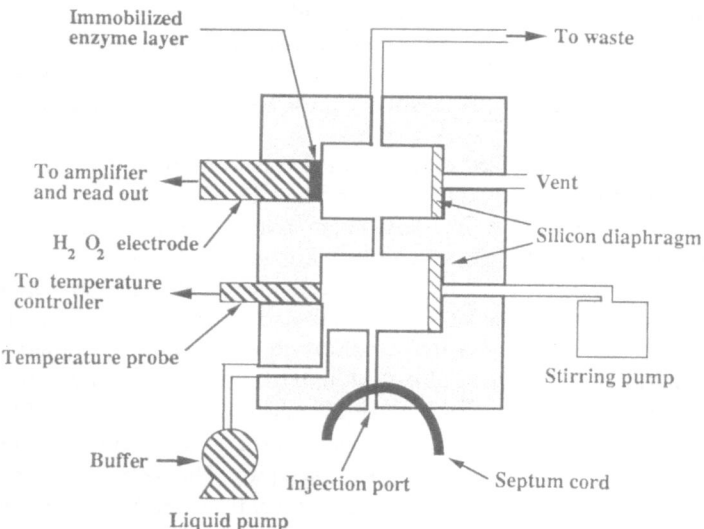

Fig. 5. Schematic diagram of a detection chamber equipped with an amperometric biosensor and other accessories

into a buffer carrier stream, it reacts with the immobilized oxidase to liberate H_2O_2, which is detected downstream (Fig. 6). Of course, the flow injection system provides better results in view of sensitivity, precision, detection limit and sampling rate, if the enzyme can be immobilized on a membrane and attached directly on the sensing area of the electrode. The theory and practice of FIA-biosensor has been recently published [38]. The FIA-biosensor systems are used for the determination of glucose [39], lactate [40], tyrosine [41], choline esterase [42], ethanol [43], uric acid [44], aspartame [45], and phosphate [46]. This list, of course, is not exhaustive as there are more oxidase-based FIA biosensors reported in literature. As compared to enzyme electrodes, the major disadvantage of FIA is its increased complexity requiring a precision pump and injection valve.

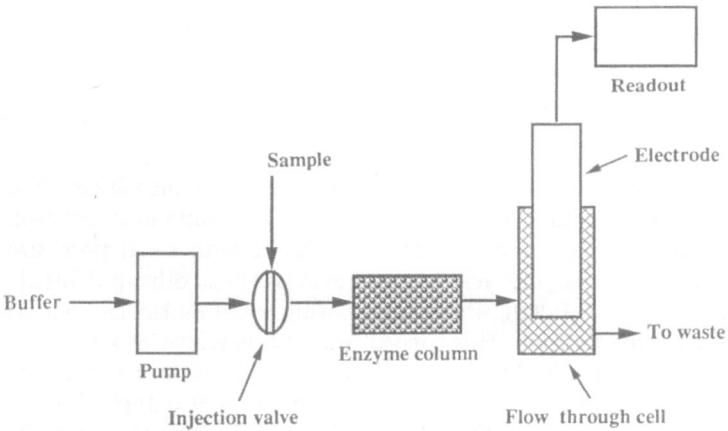

Fig. 6. Flow injection analysis system using an immobilized enzyme column and an amperometric electrode downstream

4.3 Continuous Measurement

Continuous measurements are always desirable for in vivo blood monitoring, process control, and waste water treatment. These measurements are often made by placing the biosensor in a flowing stream since in-situ monitoring would present several shortcomings. One of the most serious drawbacks of these applications is the relative instability of the enzyme which requires frequent calibration to compensate for the loss in enzyme activity. In most cases the biosensor has to be removed from the flow stream and the process of measurement is interrupted. This drawback has to be overcome before biosensors find widespread use for continuous measurement.

5 Solutions to Interference

5.1 Differential Measurement

One solution to interfering substances in the sample is the use of two identical electrodes: one with enzyme and one without enzyme [47]. Measurement is performed by substraction of current from an electrode without enzyme from current of an electrode with oxidase to eliminate the interfering current. This concept, patented by Clark [47] in 1970, formed the basis of the first commercial glucose analyzer produced by the Yellow Springs Instrument of Ohio, USA (Model 23 YSI Analyzer). The main disadvantage of this system is the requirement of an extra electrode and complicated electronic circuitry. In practice, it is somewhat difficult and/or expensive to produce electrodes which exhibit identical performance characteristics in view of response time and sensitivity.

5.2 Permselective Membrane

Another solution to interfering substances it to place a membrane such as cellulose acetate which is permeable only to low molecular weight compounds, between the enzyme layer and the electrode [47]. This membrane with small pore size (cutoff 100 Da) excludes glucose, uric acid, ascorbic acid and most other potentially interfering substances (MW > 200), while still allowing small molecules such as hydrogen peroxide to pass through. This concept was also patented by Clark and commercially exploited by the Yellow Springs Instrument for glucose and lactate measurements. The use of a H_2O_2 permselective membrane is also applied in the commercial uric acid analyzer UA 300 A (Fuji Electric, Japan) and the GKM 02 variant of the Glukometer [48]. The disadvantage of this system, however, is that it creates an additional diffusional layer which may adversely affect the sensitivity and the response time of the electrodes. It should also be noted that some blood preservatives and certain drugs do reach the platinum electrode where their oxidation produces erroneous results. The linear range of these biosensors is limited at high substrate concentrations because oxygen becomes a limiting factor. As the analyte concentration increases, more oxygen, the co-substrate is required to complete the enzymatic reaction. When the oxygen concentration is too low, each increase in analyte concentration no longer results in a linear increase in the overall reaction rate, and the biosensor response begins to level off and eventually becomes independent of analyte concentration. Another solution is to cover the enzyme layer with a negatively charged membrane for the selective retardation of anionic species [49]. This system greatly facilitates the detection of hydrogen peroxide in the presence of large amounts of endogeneous ascorbate and urate.

5.3 Anti-Interference Enzyme Layer

Another interesting approach to circumvent interfering substances is the use of an "anti-interference" enzyme layer [50, 51]. For instance, the measurement of glucose in serum may be interfered by ascorbic acid, this problem can be alleviated by covering the "anti-interference" membrane over the glucose oxidase membrane. The "anti-interference" membrane, consisting of ascorbate oxidase and catalase, converts ascorbic acid to non-disturbing compounds. Obviously, this concept can be extended to other enzyme systems such as urate oxidase/catalase for uric acid, glutathione oxidase/catalase for glutathione, etc., if the measured sample contains the aforementioned interfering suspect(s). The main drawbacks of this system, as discussed previously, are the increased thickness and number of membranes which lengthen the response and recovery times, and also necessitate more frequent calibration. The upper-linear range of these biosensors is also very limited.

The problem of interference may also be overcome using an enzyme reactor column with the hydrogen peroxide electrode downstream. The sample first bypasses the enzyme column where the background current is determined. The sample is then pumped through the enzyme column before being introduced to the electrode. The difference between these two readings is equated to the analyte concentration. An obvious disadvantage of this system is the large amount of enzyme required. Another drawback of this system is its increased complexity, requiring a pump and a flow-divided scheme. The enzyme column, however, has a long operating life and is good for several analyses.

5.4 Mediator Based Biosensors

The potentials applied to dismute (500 mV to 800 mV) are sufficiently extreme so as to introduce the possibility of interference. As discussed previously, interference can be alleviated by the use of permselective membranes which prevent the unwanted substances from reaching with the platinum surface, the use of an "anti-interference" enzyme layer, and differential measurement by using two identical electrodes: one with enzyme and one without. Among these methods, the use of the permselective membrane is a popular choice since such a membrane is more applicable and effective for samples containing several interfering species. A potential as low as 400 mV can be applied to a platinized carbon electrode vs SCE for the detection of H_2O_2. A significant reduction in the potential applied to such an electrode system would, therefore, enhance the selectivity of the platinized carbon electrode [52].

Some of the difficulties can be overcome by the use of mediators. These compounds replace oxygen to effect the electron transfer from the enzyme prosthetic group to the electrode, in order to restore the enzyme to the resting state, so that it is ready for further catalytic action.

At first glance, it appears that the use of mediators will automatically overcome the problem of oxygen dependence, and that the effect of pH will be eliminated since the oxidation of H_2O_2 does not take place and therefore protons are not

involved. However, these expectations will materialize only if the following criteria are also met:
1. The regeneration of the oxidized mediator is pH independent;
2. The reduced mediator does not react with oxygen;
3. The reduced enzyme prosthetic group reacts with the mediator much faster than with oxygen.
Apparently, these criteria are not simultaneously satisfied and the oxygen dependence is still a problem with many mediated biosensors.

It can be expected that the problem of interference by electroactive species is also alleviated with the use of mediators, since they can generally be generated at low potentials (< 400 mV). Various chemicals including quinone, organic ions and inorganic ions have been tested for mediating properties. Ferrocene (η^5-bis-cyclopentadienyl iron) has attracted the most attention because it can be easily derivatized [53]. A wide variety of ferrocene derivatives have been synthesized and studied, and several of them have been found to closely match the above-mentioned criteria. In one case, a mediator permitted the use of an applied potential as low as 50 mV [54]. However, the use of a mediator is an added complexity since it must be introduced with the sample at the time of analysis or immobilized with the enzyme. The introduction at analysis time presents an inconvenience which is hardly acceptable, while the stability of an immobilized mediator must be fully assured for the mediated biosensor to be of practical value. In addition, leakage appeared to be a severe problem, as should be expected due to the small size of the mediators. A conflicting requirement also seems to be at play. If a mediator is expected to act as an electron shuttle, it must be free to diffuse between the enzyme prosthetic sites to the electrode metal. A strong anchorage, in an attempt to prevent leakage, might negate the usefulness of a mediator. In the particular case of glucose oxidase which is known to possess a thick glycoprotein shell, and deeply embedded prosthetic groups, non diffusing mediators are not expected to be very effective. Since mediated sensors are envisioned for in vivo applications, the leakage problem raises the question of toxicity. Of course, if these hurdles are not satisfactorily surmounted the sensors will have very little practical value.

One possible solution to the problem of mediator retention and mediator efficiency is the incorporation of mediator molecules into the enzyme structure. This approach appears logical but has not been well studied and developed [55]. The foregoing, are merely some excursive discussions on mediated sensors, an excellent overview on the development of mediated biosensors is available and should be consulted for detailed information [56].

Despite the expected complication, several mediated sensors have been described, again glucose oxidase has been the model although L-amino acid oxidase [57], lactate oxidase [58], and xanthine oxidase [59] have also been studied. It is of particular interest to note that a potentially implantable glucose sensor with mediated electron transfer having satisfactorily operating stability has been reported [60]. Apparently, the exploitation of mediators will culminate in implantable microsensors in the near future [61].

6 Applications of Hydrogen Peroxide Based Biosensors

6.1 Single Enzyme Biosensor System

The classical example of hydrogen peroxide based biosensors is the development of a glucose biosensor using glucose oxidase. Immobilized glucose oxidase used with a hydrogen peroxide sensor has now been commercialized by Yellow Springs Instrument Co. (USA), Fuji Electric Co. (Japan), Kyoto Daichi Kagaku (Japan), Omron Toyoba (Japan), Solea-Tacussel (France), VEB-MLW Prufgerate-Medigen (FRG) [6]. Commercial biosensors are also available for other clinically important substances such as lactate and urate using immobilized lactate oxidase (Yellow Springs Instrument and Omron Toyoba) and urate oxidase (Fuji Electric Co.), respectively, [6]. Several biosensors exploiting the detection of hydrogen peroxide produced by the action of oxidase have been developed. These include pyruvate oxidase [62], lactate oxidase [63], sulfite oxidase [64], xanthine oxidase [65], cholesterol oxidase [66], glutamate oxidase [67], galactose oxidase [68], alcohol oxidase [69], L-amino acid oxidase [70], and choline oxidase [71]. This list, of course, is not meant to be exhaustive, but only to illustrate the potential application of the single enzyme biosensor system.

6.2 Multi-Enzyme Biosensor Systems

The detection of hydrogen peroxide has also become an established approach to the development of multienzyme electrode systems. The use of multiple enzymes is necessary under one of the following circumstances: (a) measurement of a particular analyte cannot be achieved with a single enzyme, (b) using multi-enzyme systems for substrate recycling to increase the sensitivity of a biosensor, and (c) use of multiple enzymes in anti-interference systems.

6.2.1 Multi-Enzyme Systems for a Target Analyte

In several applications, the measurement of a target analyte cannot be realized with a single oxidase enzyme. In this case, one or more enzymes are used together with the oxidase enzyme to form the biosensor. One example is the construction of a maltose biosensor by immobilizing glucoamylase on one side of a collagen membrane and glucose oxidase on the other [72, 73]. The glucoamylase converts the maltose to glucose, which diffuses through the collagen membrane and reacts with the glucose oxidase producing hydrogen peroxide, which is then oxidized at the platinum electrode as described previously. Following this approach, several biosensors have been developed. These include invertase and mutarotase and glucose oxidase for sucrose [74], β-galactosidase and glucose oxidase for lactose [75], D−phospholipase and choline oxidase for phosphatidyl choline [76], cholesterol ester hydrolase (esterase) and cholesterol oxidase [77] for total cholesterol, creatine aminohydrolase and sarcosine oxidase for creatine [78], creatinine amino hydrolase and creatine aminohydrolase and sarcosine oxidase for creatinine [78], nucleoside phosphorylase and xanthine oxidase for inosine [27] or phosphate [46].

peptidase and aspartate aminotransferase and glutamate oxidase for aspartame [65], glutaminase and glutamate oxidase for glutamine [79]. Again, this list is not meant to be complete but only illustrates some possible applications of multienzyme biosensor systems. A new multienzyme biosensor system can be constructed for an array of analytes provided a suitable coupled reaction is available. For instance, the enzyme guanine aminohydrolase can be used together with xanthine oxidase to form a biosensor for guanine; since the first converts guanine to xanthine and the latter oxidizes xanthine to uric acid and hydrogen peroxide which can be determined amperometrically.

6.2.2 Substrate Recycling

Amperometric detection coupled with substrate recycling presents a novel use of multienzyme systems. The method utilizes two different enzymes, E_1 (oxidase) and E_2, working in conjunction. The original substrate (S) is converted to product P, which is in-turn recycled to S. M and/or M* is an electroactive detectable compound (Fig. 7). One of the best know systems is the recycle of L-lactate using lactate oxidase (LOD) and lactate dehydrogenase (LDH) [80].

$$\text{L-lactate} + O_2 \xrightarrow{\text{LOD}} \text{pyruvate} + H_2O_2 \tag{6}$$

$$\text{pyruvate} + \text{NADH} \xrightarrow{\text{LDH}} \text{L-lactate} + \text{NAD}^+ \tag{7}$$

Another interesting system is the cyclic substrate conversion using glutamate oxidase (GLOX) and glutamate dehydrogenase (GLDH) [81].

$$\text{L-glutamate} + O_2 \xrightarrow{\text{GLOX}} \alpha\text{-ketoglutarate} + NH_3 + H_2O_2 \tag{8}$$

$$\alpha\text{-ketoglutarate} + \text{NADH} \xrightarrow{\text{GLDH}} \text{glutamate} + \text{NAD}^+ \tag{9}$$

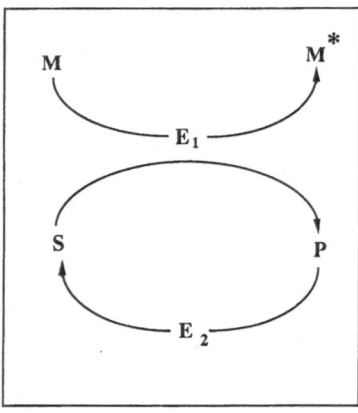

Fig. 7. Schematic diagram of a substrate recycling system

The main advantage of substrate recycling is that the system allows the determination of analytes that are not accessible with only an oxidase enzyme. Also, this system often provides a higher sensitivity, since the targeted substance is recycled several times to enable the production of H_2O_2 beyond the stoichiometric limitation. An obvious drawback of this system is the supply of $NAD(P)^+/NAD(P)H$, an expensive and unstable co-factor as described earlier. However, a new substrate recycling system was recently reported for glutamate using glutamic oxaloacetic transaminase (GOT) instead of glutamate dehydrogenase [82].

$$\alpha\text{-ketoglutarate} + \text{L-aspartic acid} \xrightarrow{\text{GOT}} \text{L-glutamic acid} + \text{oxaloacetate}. \tag{10}$$

This concept is yet to be demonstrated in a biosensor configuration for the determination of glutamate.

6.2.3 Anti-Interference System

An interesting use of the multienzyme system is to reduce the level of interfering substances in the measured sample. Laccase was co-immobilized with glucose oxidase in a gelatine membrane to form a biosensor for glucose [50]. $Fe(CN)_6^{3-}$ was added to the samples to oxidize reductive interfering substances such as ascorbic acid. The $Fe(CN)_6^{4-}$ formed is reoxidized by a laccase-catalyzed reaction. This method was reported to eliminate ascorbic acid up to 20 mM in the sample.

6.2.4 Critique on Multienzyme Systems

There are several limitations in establishing the overall optimal operating conditions in view of pH, temperature, type of buffers, stabilizers, etc., for the multienzyme system. Optimal conditions for one enzyme are not often ideal for another enzyme or enzymes, and a compromise must be made in the operating conditions. The overall stability of the biosensor system is dictated by the least stable enzyme and the measurement procedure becomes more complicated if the sample contains both the targeted analyte and substrate for the oxidase enzyme. For instance, the determination of sucrose using the "sucrose membrane" is subject to interference by the glucose content of the sample because the "sucrose membrane" consists of invertase, mutarotase and glucose oxidase. In such a sample, glucose must also be determined using the glucose membrane only (glucose oxidase), in addition to the sum of glucose and sucrose. This procedure appears to be not very convenient, since the membranes have to be exchanged during the measurement. The problem of glucose interference in sucrose determination can be partially solved by covering the "sucrose membrane" with an "anti-interference" layer consisting of glucose oxidase and catalase. β-D-Glucose permeating onto the anti-interference layer is converted to non-disturbing products such as gluconic acid, O_2, and water by the combined action of glucose oxidase and catalase. Using this anti-interference layer, determination of sucrose was reported to be unaffected by endogenous glucose, if the glucose concentration does not exceed 2 mM. Of course, this concept can be extended to other multienzyme biosensor systems. A drawback of this technique is its complexity as discussed earlier. The thicker and more complicated the biochemical layer and its associated membranes, the more difficult it is to maintain constant diffusion of reactants and products, thus increasing the frequency of calibration. As the membrane layers become thicker, the response and recovery times are expected to increase accordingly. The upper-linear range is somewhat limited and this problem is more pronounced for application in anoxic or anaerobic situations due to the lack of oxygen.

6.3 Amperometric Sensor for Immunoassays

The hydrogen peroxide sensor has also been used for immunological assays [83, 84]. The simplest approach is to immobilize an antibody onto a membrane and place the resulting membrane on the surface of the electrode. The oxidase labelled antigen and sample antigen competitively react with the immunosensor. After the reaction is completed, the sensor is washed and added to a solution containing enzyme substrate. The unknown concentration of sample antigen is related to the activity of the labelled enzyme measured by the electrode. At present, the sensitivity of this technique is considerably low when compared with established enzyme immunoassays and a new membrane must be used for each assay. It is still problematical to dissociate labelled and/or sample antigen from the bound antibody, since the antibody-antigen interaction is very strong and irreversible. Another drawback of this system is that the measurement is too time-consuming. The whole incubation time required for a complete interaction between antigen-antibody can be several hours. A general solution to this problem has been to immobilize the antibody, onto water insoluble particles or on the internal wall of a coil to form an immunoreactor. When the incubation is completed, the substrate of the labelled enzyme is added and the detection of hydrogen peroxide is followed.

A sandwich-type assay using a hydrogen peroxide electrode is another possibility. Similarly, an antibody is immobilized on a membrane or solid particles to form an immunoreactor. The targeted antigen is allowed to interact upon the immobilized antibody to form the antibody-antigen complex. The oxidase labelled antibody then reacts with the antigen of the antibody-antigen complex to form the antibody/antigen/labelled antibody sandwich. The addition of the enzyme substrate liberates H_2O_2, which is then detected amperometrically. Following this manner, a flow immunoassay system was developed for the detection of *Salmonella typhimurium* by immobilization of antibody to *Salmonella* onto polyethylene tubing by covalent binding using glucose oxidase as marker [85]. Based on the sandwich assay, the antigen was quantified by electrochemical detection of H_2O_2 produced from the enzymatic oxidation of glucose to gluconic acid. The immunoreactor could be regenerated for at least 50 times during one month of the experiment and its sensitivity was comparable to that of ELISA assay ($10^5 - 10^6$ cells per ml).

7 Critique on Biosensors Based on Hydrogen Peroxide Measurement

As described previously, amperometric biosensors are relatively simple in application and preparation. They possess: a low detection limit, a large dynamic range and good reproducibility. Together with oxidase systems, amperometric electrodes based on hydrogen peroxide measurement continue to have a dominant position in both research and commercialization. Although functional, there are a number of disadvantages to biosensors using oxidase enzymes and a hydrogen peroxide detector:

1. The hydrogen peroxide produced may inactivate the enzyme, i.e., reducing the operating life of the enzymic membrane.
2. The reaction product(s) may impede the performance of the hydrogen peroxide electrode. For instance, phenol, when oxidized, produces a black shiny deposit on the electrode surface which alters its chemical reactivity [23].
3. The electrode is dependent on the oxygen availability in the system since O_2 is a co-substrate of oxidase catalyzed reactions. Therefore, the electrode ceases to function in anoxic or anaerobic situations. For applications in which the concentration of oxygen is much lower than the K_M values of the oxidase enzyme, the output of the biosensor might be nonlinearly proportional to the substrate concentration.
4. The electrode response varies quite considerably with the pH of the solution since the oxidation of H_2O_2 to oxygen is proton dependent ($H_2O_2 \rightarrow 2\,H^+ + O_2 + 2\,e^-$). For reliable application of the hydrogen peroxide electrode, both pH and pO_2 of the solution must be carefully controlled.
5. The relative lack of long term stability of the oxidase system is the most serious drawback of biosensors in on-line and/or continuous monitoring. To identify the causes of activity loss is still empirical, as well as by trial and error. The low activity of some oxidase systems is another matter of concern. Chemical modification of enzymes to improve their activity has been reported with limited success. Modification of the lysine residues of alcohol dehydrogenase with methylpicolinimidate increased the enzyme's activity 19 times [86]. To date, except for glucose oxidase, none of the important oxidases used in biosensor construction is being subjected to chemical modifications. Of course, the prerequisite for such a task is to understand the enzyme structure and enzyme-substrate interaction. Unfortunately, such necessary background information is not often available for the particular enzyme of interest. For instance, there is no published amino acid sequence for glucose oxidase, the most widely used enzyme, even though substantial quantities of this enzyme are available in a highly purified form [86] and crystallization and X-ray diffraction studies of a deglycosylated glucose oxidase were reported recently [87]. The engineering of enzymes for biosensors is still in its infancy, however, this technique becomes important in the next generation of biosensor devices. Some potential targets for protein engineering include: (a) shift in or removal of pH dependence, (b) change in linear response range to substrate concentration, (c) improved stability during storage and operation, (d) reduction in

susceptibility to interfering substances, and (e) widening or narrowing of substrate specificity. For instance, glucose oxidase could be modified by reacting with sodium periodate to oxidize its carbohydrate residues to carbonyl groups. Such a treatment led to improvement in both sensitivity and stability of the enzyme [88]. Glucose oxidase could also be modified by crosslinking with acid phosphatase, invertase, peroxidase and glucosyl transferase. Crosslinking modification of glucose oxidase led to pronounced increase in stability against denaturation and proteolysis [89].

8 Conclusions

Together with oxidase systems, amperometric electrodes based on hydrogen peroxide measurements continue to have a dominant position in research and development, and commercialization. The main advantage of hydrogen peroxide over the oxygen electrode based biosensor is its excellent sensitivity. Recently, flow injection techniques have been developed to improve sensor capability in terms of the detection limit, stability, and reproducibility. The major drawback of the hydrogen peroxide based biosensors, however, is the poor selectivity of the electrode. The measurement of the targeted analyte in samples containing amperometrically active interfering substances such as uric acid, ascorbic acid, glutathione, cysteine, acetaminophen, bilirubin, tyrosine, etc., is still problematical. Although, this problem has been reduced significantly with the use of a reference electrode, a permselective membrane, and/or an anti-interference membrane, it has not been totally eliminated.

To overcome the problem with interference, modified and mediated amperometric biosensors have received much attention in recent years. Ferrocene and its derivatives appear to be the best candidates even though other mediators such as tetrathiafulvalence-tetracyanoquino dimethane (TTF-TCNQ) have also been used widely in biosensor construction. The search will be continued for an ideal mediator in view of low redox potential and fast transfer rates between the mediator with both the enzyme and the electrode. Technical "know-how" to confine the mediator to the enzyme layer and prevent its leakage must also be better developed in a more practical way. This is a crucial research area which requires immediate attention before mediated amperometric biosensors can become practical.

From a biological viewpoint, the relative lack of long term stability of the oxidase system is the most serious limitation of biosensors. The problem of instability can be circumvented by using enzymes stable at high temperatures, naturally available from thermophilic microorganisms [90]. The limitation of this approach is the very slow reaction rate of such enzymes at room temperature. As the structure of enzymes becomes better defined, it will be possible to "tailor" enzymes that can function in a specific way and with more stability. Chemical modification of oxidase enzymes to make them more stable and more specific is another interesting approach. Modified enzymes are likely to become important in the next generation of biosensors. At present, none of the important enzymes in biosensor development is modified by protein engineering.

In the past, microorganisms and their components were only exploited to make products such as beer and bread. In the modern era of biotechnology, biological materials can be used for biosensor construction. Biosensors will have an impact on people's daily lives in the not too distant future, since they can fill real needs in health care, environmental monitoring and control, agriculture and chemical/biochemical industries.

9 References

1. Clark LC Jr, Lyons C (1962) Ann NY Acad Sci 102: 29
2. Kadish AH, Hall DA (1965) Clin Chem 11: 869
3. Updike SJ, Hicks GP (1967) Nature 214: 986
4. Guilbault GG, Luong JHT (1989) Selective Electrode Reviews 11: 3
5. Turner APF (1987) Preface. In: Turner APF, Karube I, Wilson GS (eds) Biosensors fundamentals and applications. Oxford Science Publications, Oxford, p.v
6. Luong JHT, Mulchandani A, Guilbault GG (1988) Tr Biotechnol 6: 310
7. Coughlan MP, Kierstan MPJ, Border PM, Turner APF (1988) J Microbiol Methods 8: 1
8. Dixon M, Webb CE (1979) Enzymes. Academic Press, New York
9. Keilin D, Hartree EF (1952) Biochem J 50: 331
10. Dixon M (1938) Enzymologia 5: 198
11. Janssen FW, Ruelius HW (1968) Biochim Biophys Acta 151: 330
12. Guilbault GG, Sadar S (1969) Anal Letters 2: 41
13. Nanjo M, Guilbault GG (1975) Anal Chim Acta 75: 169
14. Chibata I (1978) Immobilized enzymes, Wiley, New York
15. Guilbault GG, de Olivera-Neto G (1985) Immobilized enzyme electrodes. In: Woodward J (ed) Immobilized cells and enzymes: A Practical Approach. IRL Press, Oxford, pp 55–74
16. Scheller F, Schubert F (1989) Biosensoren, Birkhäuser, Basel p 17
17. Wilson GS (1987) Fundamentals of amperometric sensors. In: Turner APF, Karube I, Wilson GS (eds) Biosensors fundamentals and applications, Oxford, pp 165–179
18. Willard HH (1974) Instrumental methods of analysis. D Van Nostrand, New York, p 646
19. Guilbault GG (1976) Handbook of enzymatic methods of analysis, Marcel Dekker, New York
20. Zaborsky OR (1973) Immobilized Enzymes. CRC Press, Ohio
21. Weetall HH (1975) Immobilization by covalent attachment and by entrapment. In: Messing RA (ed) Immobilized enzymes for industrial reactors. Academic Press, New York, pp 99–123
22. Scouten WH (1987) A survey of enzyme coupling techniques. In: Mosbach K (ed) Methods in Enzymol. 31: 135, pp 30–65
23. Clark LC Jr (1987) The Enzyme electrode. In: Turner APF, Karube I, Wilson GS (eds) Biosensors fundamentals and applications. Oxford Science Publications, Oxford, pp 1–12
24. Guilbault GG (1984) Analytical uses of immobilized enzymes. Marcel Dekker, New York
25. Mulchandani A, Luong JHT, Male KB (1989) Anal Chim Acta 221: 215
26. Coulet PR, Julliard J, Gautheron DC (1988) French Patent 73-23283
27. Mulchandani A, Male KB, Luong JHT (1990) Biotech Bioeng 35: 739
28. Bélanger D, Nadeau J, Fortier G (1989) J Electroanal Chem 274: 143
29. Foulds NC, Lowe CR (1988) Anal Chem 60: 2473
30. Deshpande MV, Hall EAH (1990) Biosensors Bioelectronics 5: 431
31. de Taxis du Poët P, Miyamoto S, Murakawi T, Kimura J, Karube I (1990) Anal Chim Acta 235: 225
32. Narasimhan K, Wingard LB Jr (1986) Anal Chem 58: 2984
33. Gregg BA, Heller A (1990) Anal Chem 62: 258

34. Tsuji I, Eguchi H, Yasukauchi K, Unoki M, Taniguchi I (1990) Biosensors Bioelectronics 5: 87
35. Degani Y, Heller A (1988) J Phys Chem 110: 2615
36. Geise RJ, Yacynych AM (1989) Electropolymerized films in the construction of bio-sensors. In: Murray RC, Dessy RE, Heineman WR, Janata J, Seitz NR (eds) Chemical sensors and microinstrumentation ACS Symp Series 403: 65
37. Nguyen AL, Luong JHT, Yacynych AM (1991) Biotech Bioeng 37: 729
38. Olsson B, Lundbaeck H, Johansson G, Scheller F, Nentwig J (1986) Anal Chem 58: 1046
39. Wiek HJ, Heider GH, Yacynych AM (1984) Anal Chim Acta 158, 137
40. Scheller F, Schubert F (1986) Anal Letters 19: 1691
41. Pacakova V, Stulik K, Brabcova P, Barthova J (1984) Anal Chim Acta 159: 71
42. Yao T (1983) Anal Chim Acta 159: 71
43. Blaedel WJ, Wang J (1980) Anal Chem 52: 1426
44. Jaenchen M, Gruenig G, Bertermann K (1985) Anal Letters 18: 1799
45. Male KB, Luong JHT, Mulchandani A (1991) Biosensors Bioelectronics 6: 117
46. Male KB, Luong JHT (1991) Biosensors Bioelectronics 6: 581
47. Clark LC Jr (1970) US Patent 3 539 455
48. Tsuchida T, Yoda K (1981) Enz Microbial Technol 3: 326
49. Lobel E, Rishpon J (1981) Anal Chem 53: 51
50. Wollenberger U, Scheller F, Pfeiffer D, Bogdanovskaya VA, Tarasevich MR, Hanke G (1986) Anal Chim Acta 187: 39
51. Scheller F, Renneberg R (1983) Anal Chim Acta 152: 265
52. Benneto HP, Deckeyzer DR, Delaney GM, Koshy A, Mason JR, Razak LA, Stirling JL, Thurston CF (1987) Int Analyst 8: 22
53. Cass AEG, Davis G, Francis GD, Hill HAO, Aston WJ, Higgins IJ, Plotkin EV, Scott LDL, Turner APF (1984) Anal Chem 56: 667
54. Johsson G, Gorton L (1985) Biosensors 1: 355
55. Degani Y, Heller A (1987) J Physical Chem 91: 1285 and (1988) J Am Chem Soc 110: 2615
56. Cardosi MF, Turner APF (1987) The relation of electron transfer from biological molecules to electrodes. In: Turner APF, Karube I, Wilson GS (eds) Biosensors funda-mentals and applications. Oxford Science Publications, Oxford, pp 257–275
57. Dicks JM, Aston WJ, Davis S, Turner APF (1986) Anal Chim Acta 182:103
58. Prenta AZ (1984) Studies on lactate oxidizing enzymes and their applications to ferrocene based enzyme electrodes for lactate, Ph.D Thesis, Cranfield Inst Technol, Cranfield, Bedford, UK
59. Cenas NK, Poncius AK, Kulys JJ (1983) Bioelectrochem Bioeng. 11: 61
60. Pickup JC, Shaw GW, Claremont DJ (1989) Biosensors 4: 109
61. Claremont DJ, Pickup JC (1987) In vivo chemical sensors and biosensors in clinical medicine. In: Turner APF, Karube I, Wilson GS (eds) Biosensors fundamentals and applications, Oxford Science publications, Oxford pp 356–376
62. Kihara K, Yasukawa E, Hayashi M, Hirose S (1984) Anal Chim Acta 159: 81
63. Mullen WH, Churchouse SJ, Keedy FH, Vadgama PM (1986) Clin Chim Acta 157:191
64. Fonong T (1986) Anal Chim Acta 184: 287
65. Ianniello RM, Lindsay TJ, Yacynych AM (1982) Anal Chem 54: 1098
66. Bertrand C, Coulet PR, Gautheron DC (1979) Anal Letters 12: 1477
67. Yamauchi H, Kusakabe H, Midorikawa Y, Fujishima T, Kuninaka A (1984) Enzyme electrode for specific determination of L-glutamate, in: Eur Congr Biotechnol 3rd, Verlag Chemie, Weinheim p 705
68. Taylor PJ, Kmetec E, Johnson JM (1977) Anal Chem 49: 789
69. Guilbault GG, Lubrano GJ (1974) Anal Chim Acta 69: 189
70. Guilbault GG, Lubrano GJ (1974) Anal Chim Acta 69: 1983
71. Mascini M, Moscone D (1986) Anal Chim Acta 179: 439
72. Coulet PR, Bertrand C (1979) Anal Letters 12: 581
73. Renneberg R, Scheller F, Riedel K, Litschko E, Richter M (1983) Anal Letters 16(B12): 877
74. Scheller F, Karsten Ch (1983) Anal Chim Acta 155: 29

75. Cordonnier M, Lawny F, Chapot D, Thomas D (1975) FEBS Letters 592: 263
76. Karube I, Hara K, Satoh I, Suzuki S (1979) Anal Chim Acta 106: 243
77. Wollenberger V, Kuhn M, Scheller F, Deppmeyer V, Janchen M (1983) Bioelectrochem Bioeng 11: 307
78. Tsuchida T, Yoda K (1983) Clin Chem 29: 51
79. Wollenberger U, Scheller F, Bohmer A, Passarge M, Muller HG (1989) Biosensors 4: 381
80. Mizutani F, Yamanaka T, Tanabe Y, Tsuda K (1985) Anal Chim Acta 177: 153
81. Schubert F, Kirstein D, Scheller F, Appelqvist R, Gorton L, Johansson G (1986) Anal Letters 19: 1273
82. Luong JHT, Mulchandani A, Male KB (1991) Enz Microbial Technol 13: 116
83. Karube I, Suzuki M (1986) Biosensors 2: 343
84. Aizawa M (1987) Philos Trans Royal Soc London B, 316: 121
85. Luong JHT, Prusak E (1990) Anal Letters 23/10: 1809
86. Cass AEG, Kenny E (1987) Protein engineering and its potential application to biosensors. In: Turner APF, Karube I, Wilson GS (eds) Biosensors fundamentals and applications. Oxford Science Publications, Oxford, pp 113–132
87. Kalisz HM, Hecht H-J, Schomburg D, Schmid RD (1990) J Mol Biol 213: 207
88. Brooks SL, Ashby RE, Turner APF (1987/88) Biosensors 3: 45
89. Barbaric S, Kozulic B, Lenstek I, Pavlovic B, Cesi V, Mildner P (1984) Cross-linking of glucoenzymes via their carbohydrate chains. In: Third European Congress on Biotechnology, Vol I, VCH, Weinheim pp 307–312
90. Twork JV, Yacynych AM (1986) Biotechnol Progress 2: 67

Author Index Volumes 1–50

Subject Index